ENVIRONMENTAL CONSERVATION: CHALLENGES & ACTIONS

By

Asish Ghosh, Ph.D

Centre for Environment & Development
329, Jodhpur Park, Calcutta 700068

A.P.H. PUBLISHING CORPORATION
4435-36/7, ANSARI ROAD, DARYA GANJ
NEW DELHI-110 002

Published by
S.B. Nangia
A P H Publishing Corporation
4435-36/7, Ansari Road, Daryaganj
New Delhi 110002
Ph.: 23274050
E-mail : aphbooks@gmail.com

ISBN - 978-81-313-0423-5

2023

Rs. 2995/-

Printed at
Balaji Offset
Navin Shahdara, Delhi 110032

PREFACE

The World at large perhaps looked for the first time at the very basic question – "Development at what cost?" in June 1972, at the UN Conference on Human Environment and Development. That the level of global concerns changed dramatically within 20 years was witnessed at UN Conference on Environment and Development held in Rio de Janeiro, Brazil, 1992. More than 180 countries participated in the meet and Agenda for 21st Century was decided. The "Millennium Development Goals" 2000 and UN Summit on Sustainable Development in 2002 made the urgency of the tasks more clear. But now at the nearing end of first decade of 21st Century, problems still persist, specially in developing countries while tangible solutions appear elusive.

In the present publication, I have tried to select 20 topics of environmental concerns and have made attempts to analyze the current situations. While actions taken can be apparently praiseworthy new thoughts have emerged regarding impacts of such action (example growing plants for Biofuel in degraded land while land remains a finite resource at the cost of rising food price). New challenges have emerged seeking new thoughts for action.

The question is how much sensitized the civil society has become in India? While public participation in decision making is a proclaimed policy in the country, the reality speaks otherwise. A series of national policies (e.g. Agriculture, Water, Population, Forest, Environment etc.) have been changed during the post globalization period but in the process some of the real issues have been sent to the back seat. The country still lacks a uniform land policy as a result of which several incidents of conflicts find place in the media. So is the case of implementing the policy of promoting renewable and new energy initiative. The forest policy still remains silent on the issue of determining net present value (NPV), while an alarming increase of conversion of forests for non-forestry purposes can be noted. In all such issues, the civil society can play a positive role. The present publication is expected to empower the society at large; however limited may be its scope, with information's of the happenings of the current times and demand environmental security while moving onwards to the path of development.

March, 2008

A K Ghosh, Ph. D.
Calcutta

CONTENTS

	Preface	*v*
1.	Human Society: Perspectives and Development	1
2.	Poverty and Development	10
3.	Environment and Human Rights	20
4.	Right to Information: Environment and Quality of Life	27
5.	Land Resources: Need for a Policy	34
6.	Land Resource: Poverty, Inequality and Rights	42
7.	Water Crisis	53
8.	Biodiversity and Climate Change	61
9.	Future Biodiversity Scenarios in India & South-East Asia	70
10.	Sacred Groves	84
11.	Wildlife Crisis	92
12.	Invasive Species	99
13.	Global Warming	112
14.	Energy, Environment and Development	151
15.	Nuclear Power: Policy and Global Politics	170
16.	Bio-Fuel	181
17.	Looking for Alternatives: Biopesticides	198
18.	Greening of Indian Cities	205
19.	Sustainability and Corporate Governance	214
20.	Protecting the Environment: Three Decades of Indian Progress	226
	Subject Index	249

CHAPTER I

HUMAN SOCIETY: PERSPECTIVES AND DEVELOPMENT

1.0 INTRODUCTION

The human society over last 10,000 years of recorded history, however fragmented it might appear, has come a long way from the days of hunter-gatherers to pastoralists and settled agriculture. The industrial revolution, hardly 300 years old has ushered a quantum change in the use of nature, natural resources and environment. The last 100 years of history indicates fastest development mode in the arena of science and technology, in our understanding of physical laws, ecosystem dynamics, information technology, outer space science, nuclear science and myriads of new frontiers. The last century also witnessed a phenomenal growth of human population.

The world population increased from 0.75 billion in 1750 to 1.6 billion in 1900 but then in next 65 years it reached 3.3 billion and is almost sure to touch 7.0 billion mark by 2010. The period of doubling the population drastically changed in last 100 years. Per capita consumption of natural resources directly increased with the rising population. Economic growth, with an objective of material well being of human society; when well being is related with 'flows of income, consumption and employment the concerns for conventional economic theory .... have centred around problem of production (Sengupta, 2004).

This demand for accelerated production leads to conversion of living and non-living natural resources into products with the help of human labour and man-made capital. The concept of unlimited growth in a changing global order may be examined first.

2.0 ON LIMITS TO GROWTH

In 1972, a study report entitled 'Limits to Growth' was published as a World Model, supported by an intellectual group, the 'Club of Rome'. One has to remember that in the same year, the first UN Conference on Environment and Human Development took place at Stockholm, Sweden. The world at large was increasingly becoming concerned with the economic growth model; the "Club of Rome's" report showed that population and industrial output as also food production may continue to grow but the resource base is apt to decline at a faster pace and global pollution load will continue to increase – cumulative effects of such events can only lead to an

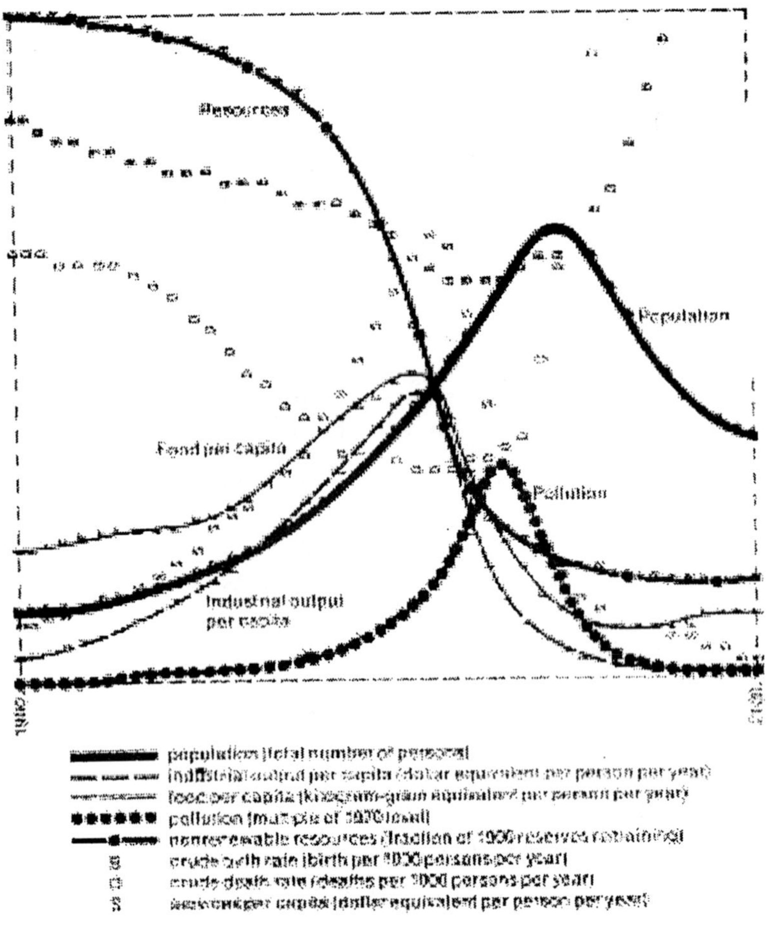

ultimate collapse of development trend. Meadows (1972) the author of the report along with others, even predicted that if no major change occurs in the present system, population and industrial growth will certainly cease 'within the next Century at the latest'. Looking back one can say after nearly one third of the period, no such indication is visible. The question is why?

The reasons however can be well evidenced. The development planners and economists were given a different signal to ensure a process of sustainability. William Nordhaus (1973) proposed a continuing expansion path over the next century, even with continued growth of World's population; this is based on assumption that 'most resources are not irretrievably lost, scarce resources can typically be replaced by substitutes and some increased output emerges from improved technology without any increase in input. This however does not justify that efforts on conservation of natural resources should not be made. Barry Commoner (1972), a well-known environmentalist, wrote during the same period that 'present course of environmental degradation, at least in the industrialized countries, represents a challenge to essential ecological system that is so serious that if continued, it will destroy the capability of the environment to support a reasonably civilized human society'. Improvements of technology perhaps help to get more from less. The increase in output per unit of input is the key in such a model. Justification of material change (say from Diesel to CNG) is another approach of looking at the way out. But Commoners' warning signal supported by serious researchers in the field of environment and economics also paved way for a new discipline i.e., Environmental Economics or Ecological Economics.

The new understanding about goods and services of nature led to the economic valuation of ecological services and assets developed over last 30 years. While the human society did not yet face a catastrophic doomsday scenario, threats are becoming increasingly perceptible. And this led to yet another document of human bondage in the form of 'Our Common Future' – an outcome of a study headed by Gro Bruntland of Norway.

3.0 IMPACTS ON ENVIRONMENT : UNEQUAL WORLD

"The rate at which humans are altering the environment, the extent of that alteration and the consequences of these changes for biological diversity are unprecedented in human history, and are now beginning to pose substantial threats to the economic and cultural life of many societies" (Watson et al., 1995). Its not only on biological diversity but also on every component of biosphere – the air, water, soil, forests, fisheries and agriculture – human impacts are causing greatest concern.

Regionally such impacts have become mere visible. Say in India, all the 4 major rivers of India have now been declared polluted. The air quality of 10 major metropolitan cities shows variable level of pollution load all above permissible limit. The vast tracts of wasteland in the country caused by mining, unsustainable agriculture and loss of land cover now amounts to nearly 30 percent of the country's land area. Good canopy forests now exist in almost 11-14 percent of total forest cover. Agricultural practices with high yielding varieties of seed, chemical fertilizer and pesticides and canal water irrigation or groundwater based irrigation led to loss of genetic resources, altered soil nutrient cycle, increased salinization etc. Most of the wildlife – the large species and medium sized species have become rare or endangered in prime wilderness areas.

In the meanwhile India's population has reached 16% of the global population that too only in 2.4% of the global space. Water has become a scarce commodity - now a root cause of inter-state and inter country dispute.

The maladies can also be ascribed to the economic system – that of distribution of wealth and stark differences in purchasing power. This economic divide between rich and the poor within the country and in the region exemplifies yet another global order of inequity. In 1960, the 20% population of the richest countries had 30 times the incomes of the poorest 20% of the world. In 1997, the relative level rose to 74 times in spite of economic growth of the developing world over nearly four decades.

In this unequal world, the issue of impact of human population growth on environment has often been sharply focused. Whether population is a resource or liability can be debated upon, keeping in view the perspective of such a discourse. However the enormous rate of consumption in the developed world with a model of stabilized population has now been questioned in terms of global resource use. Amazonian rain forest in Costa Rica – one of the richest abode of biodiversity – is turned into beef cattle ranch for producing meat for hamburger industry in USA. The limited petrochemical resources remain incredibly cheap in US market, offering a lifestyle unreachable by millions in the developing world. Over consumption not only jeopardize global resource base, it harms the society indulging in such practices in the long run; 60% of US population is termed 'obese' as par a recent statistics.

In such a scenario "any regime of steady state economic growth would mean exponentially growing flows of resources from nature to the economy and similarly growing throw backs of wastes from the economy to the nature" (Sengupta, 2004). The last century has actually provided enough evidences towards this hypothesis.

4.0 SEARCHING ALTERNATIVES

The UN Conference on Environment and Development (UNCED) held at Rio de Janeiro, Brazil in June 1992 – 20 years after Stockholm Conference, is considered the biggest ever assembly of worlds leaders (181 countries) to discuss the issue. Over 10 days of heated arguments, the representatives tried to identify the major issues of concern for human society in the changing environment. The way to reach the objectives was paved by several international negotiations, treaties and national and regional level agreements.

Internationally, environmental concern was expressed earlier in the Ramsar Convention for conservation of wetlands as far back as in 1971. Convention on International Trades in Endangered Species of Wild flora and fauna (1973), UN Convention on the Law of the Seas (1982), Vienna Convention for Protection of Ozone Layer (1985), Montreal Protocol on Substances that deplete Ozone layer (1987), Basel Convention on Trans-boundary movement of Hazardous Wastes (1989) etc. are land marks. In 1992 and post 1992 period, UN Convention on Biological Diversity (1992), UN Framework Convention on Climate Change (1992), Kyoto Protocol on Climate Change (1997), UN Convention to Combat Desertification (1994) become major instruments to combat the crisis.

The adoption of Agenda 21 as a 21st Century Roadmap for Development integrating environmental concern at the UNCED, 1992 and setting up of UN Commission on Sustainable Development following UNCED are other noteworthy land marks in last three decades but more evidenced in last 20 years.

Regionally, the setting up of SAARC forum and South Asian Cooperative Environment Programme are noteworthy.

India has embarked upon an ambitious programme of integrating human development with environment. The newly formed Department of Environment (1980) turning into a 'Ministry of Environment and Forests" (1985) bear evidences to such efforts. Environment Action Programme, reinforced by a plethora of laws and policies in India, Bangladesh, Pakistan, Nepal, Bhutan and Sri Lanka reflect the variable level of national interest on the issues in this region. The National Environment Policy of India now in the draft form has called for public comments before being finalized; this can be seen as a democratic process of ensuring public participation in the system of governance. It has since been published in 2006.

In all these efforts developmental goals remain central focal point but with a rider for ensuring sustainability.

The sustainable development process of Human Society also tries to ensure an ecological balance on the planet earth. The economic indicators of development have now been changed into a new assessment process.

5.0 GLOBAL EFFORTS

The human society with the changing environment is also witnessing a changing economic order. Globalization is urged through a process of structural adjustment of national economic and development strategy; the aim is to ensure better growth through free flow of investment, market opportunities, new employment and ultimately poverty alleviation. Can such a process reconcile with the model of sustainable development?

Since 1990's Human Development Reports of UNDP are using non-economic indicators many of which are also indicators for a process of sustainable development. For example, percentage of population with access to safe drinking water, percentage of children under 5 suffering from malnutrition, percentage of people living above the age of 65 may be good indicator of the status of quality of life. Use of such indicators could reveal the real profile of the people rather that Gross Domestic Produces (GDP) expressed in term of monetary growth.

In order to develop a national or regional model for development, new efforts are being made to revive dying wisdom of local communities in the vast rural landscape of south east Asia and elsewhere. Traditional knowledge and practices in such areas as seed selection and exchange, tank irrigation, rainwater harvesting, watershed development, use of herbal medicines, use of unconventional items from nature for food and medicines are but few of such examples. The role of civil society organizations in bringing those knowledge into the focus of mainstream planners are commendable. This is strange at a period of globalization promoting brand equity but real in terms of happenings.

Global market mercifully did not or could not prevent such a process. Simultaneously, in the year 2000 governments under the aegis of United Nations made an unprecedented resolution through millennium declaration consisting of goals in eight identified areas; this is now well known as millennium development goals (MDG). These include (i) halving the incidence of extreme poverty and hunger, (ii) providing universal access to primary education, (iii) promoting the equality of women, (iv) reducing infant mortality by two third, (v) reducing maternal mortality by three – quarters, (vi) halting and reversing the spread of diseases like malaria and HIV/AIDS, (vii) achieving environmental sustainability and (viii) forming a global partnership for development. The governments resolved that these goals are to be met by 2015 or earlier. MDGs are people oriented, time bound and measurable and perhaps will be achievable. In September 2004 the United Nations published an assessment report on the progress of MDGs. In east and south east Asia, many country are on track to meet nearly all goals by 2015. Reservation still exists on the lack of commitments with regard to the goal (viii). Instead of targeted 0.7% of GNP to be given to the poor countries only 0.25% amounting to US $ 68.5 billion has been received during the year 2002 and 2003. The forthcoming UN conference may provide a better opportunity to examine the success and failures in MDGs implementation process. (Gadtuala, 2005)

The World summit on Sustainable Development at Johannesburg, 2002 called for a new social order of equity and resource sharing. The Secretary General of United Nations urged upon the participating countries specially the highly developed countries to commit assistance for a programme called WEHAB. This is to address the critical issues of water, energy, health, agriculture and biodiversity. During the conference at least part of the commitments came from such countries. So while the global climate change potential appears to reach a dimension beyond earlier projection, and the impacts could cause significant damage to the Human Society – the global commitment provide much hope even in a changing environment.

The political divide between North and North is becoming more apparent in some key environmental issues. USA has refused to sign Bio-Safety Protocol and ratify Biodiversity Convention and has also withdrawn from Kyoto Protocol (calling for reduction of emission of green house gases specially carbon to 5.2% below the 1990 level by 2012) while European Union strongly supported all the three global efforts.

The human society is now facing a crisis of uncertainty. It was expected that with the end of cold war, at least a part of GDP (0.7%) of all developed countries would be diverted towards environmental conservation and human development in the vast arena of developing countries. That hope has largely been belied. War and ethnic confrontations are ever increasing in the defence budget of largest economy, USA, as also in the developing countries like India. The resources that could have been well utilized to meet the basic demand for food, education, health, energy, environment and habitat improvement are now being diverted to military spending.

The problems are well defined. Apprehensions are well founded. Ways are well marked, but in reality necessary action to secure Human Society is missing.

This presentation however ends with a note of high optimism. Human Society has survived many a crisis, faced challenges of war and famine in the past, might have adopted wrong courses for development but it has realized the follies at the end. Expectedly 21st Century will at the end, offer a better perspectives for Human Development not withstanding the north-south divide, poverty and inequity, challenges posed by possible climate change and loss of genetic resources.

REFERENCES

Bruntlund, Gro, 1987. Our Common Future. Oxford University Press.

Commoner, Bary, 1972. The Closing Circle. Alfred A. Knopf, New York.

Forrester, Jery, 1971. World Dynamics. Wright Allen Press. Cambridge, Mars.

Gadtaula, Prashansa, 2005. MDGs : Miles to go. *Trade Insight,* 1(1): 13 .

Meadows, Donella, H. et. al., 1972. The Limits to Growth. A Report for the Club of Rome's Project on the Predicament of Mankind. Universe Books, New York.

Nordhaus, William, O., 1973. World Dynamics: Measurement without Data. *Economic Journal*, 83:1156-1183.

Sengupta, Ramprasad, 2004. Nature, Economy and Population: Biophysical Limits and Sustainability. p. 53-100. In Sustainable Development: Ecological, Economic, Sociocultural and Statistical Aspects. (ed. A. K. Ghosh, J. K. Ghosh, Barun Mukhopadhyat). Indian Statistical Institute.

United Nations, 1992.Agenda 21.Conference on Environment and Development,U.N., New York.

Watson, R. T. et. al., 1995. Global Biodiversity Assessment: Summary for Policy Makers. Cambridge University Press, U.K.

CHAPTER II

POVERTY AND DEVELOPMENT

1.0 MILLENNIUM DEVELOPMENT GOALS

Millennium Development Goal (MDG), launched by United Nations in 2000, identified eight target areas, the first of which is reduction of income poverty; the other goals are Primary education, Gender equality, Child mortality, Maternal health, HIV / AIDS / Malaria, Environmental sustainability and Global partnership for development. Eighth goal is not an end, as some analysts believe, but a means (Kanbur, 2005).

Development invariably is linked with growth. The seven goals set up by United Nations (UN) are not all inclusive areas needed for development, but obviously prioritization was called for; question may be raised on the comparative weightage for each of the goal set.

It is believed that 'real development' only started after Industrial Revolution beginning around 1750. Extreme poverty was reduced with application of new scientific discoveries and technological innovations. It is estimated that by the turn of 20th Century, more than five billion people of the global population of 6.5 billion, got some reprieve. However, even now in 21st Century, more than 20,000 people die every day due to lack of nutrition, safe drinking water, healthcare and other essentialities. Extreme poverty is defined with income of less than \$ 1.0 a day, and these unfortunate people are deprived of all essentialities. 'Industrial Revolution' could not provide the entire humanity freedom from poverty and hunger.

Economists have often gone into reasoning to find out such inequality. Poor governance (often code words for corruption) has been one of the much used findings of such research. Besides blaming the system of governance, racial discrimination, religious taboos, caste system, gender inequalities and sheer lack of entrepreneurship are other common attributes used to answer the reasons for current level of poverty. But historically, colonial expansionism and resultant exploitation of natural resources and use of free human labour as slaves (from the colonies) may also provide some cue for such a scenario. Welfare measures were too few to mention in terms of coverage of population and extents of means in the colonial era.

Poverty today is mostly pervading in the Afro-Asian countries and South America. Poverty stricken countries were mostly under colonial powers for variable but considerable period of time. Some Western economists believe such exploitation was the result of poverty rather than cause of it. Economic history of the countries before and after colonization may prove such conjectures wrong. While Europe and North America flourished, these colonial settlement received very little of the benefits derived from colonial countries' resources. Chasm between governed one and the governing community widened. When Afro-Asian and Latin population become finally liberated, the legacy of past governance prevailed. Each of the newly liberated countries had to carry their own cross.

2.0 GROWTH, INEQUALITY AND POVERTY

The concept that often espoused is growth is directly correlated with increase in per capita income. Distribution of real income is now being questioned. Kanbur (2005) defined the distribution of real income across individuals, correcting for price variation, non-market provision of goods and services, size of household and distribution of resource within such households. Very simply he deducted 'if the objective is to reduce poverty then obviously growth is a plus for poverty reduction and increased inequality a minus'. But does growth ensure decline in inequality, if not, the increased inequality can have a significant impact on poverty. That will nullify the hypothesis that 'Growth is Good for the Poor' (Dollar and Kraay, 2002).

India, for instance, has witnessed an impressive change in per capita income during last decade and a half but change in inequality is hard to quantify. It is therefore this lack of association between growth and inequality that poses a serious question on the path that is to be chosen for poverty reduction. It is argued by the proponents of globalization that free trade confers greater benefits 'in terms of higher productivity and growth'. The hypothesis that 'trade ratio' (which is defined as imports plus export divided by Gross Domestic Produce or GDP) indicates trends in open trade and higher growth is riddled with the question of tariffs.

Tariff barriers have often been debated in the forum of WTO. Beatty (2002) stated that 'developing countries loose about $ 100 billion a year owing to Western export-subsidies and trade barriers. For agriculture alone these amounted to $ 245 billion in 2000 – five times what the West spent on development assistance that year'. It is now well known that 'deeply poor' nations (Vide World Bank definition where people live on less than a dollar a day) are strangely the worst victim. Host of commodities in consumer markets like textile, leather, footwear arrive in rich economy of United States from 49 least-developed countries who face hardest tariff, offsetting the benefit of aid flows from the West.

An OXFAM report entitled 'Eight Broken Promises: why the WTO Isn't Working for the World's Poor', illustrating the ills of such trade barriers, proclaimed 'International trade has the potential to act as a powerful source for economic growth and poverty reduction' but the potential is invariably being lost due to irrational trade barriers. The blatant betrayal on the promise to reduce agricultural subsidies by the developed countries in exchange of opening up market for agricultural imports (by reducing to 50 per cent) literally led to awe and shock. But Western countries still merrily impose tariff on import from developing countries 'as high as five times as these on all other goods'. In EU countries meat and diary imports may face a tariff of more than 100 per cent. Logically the benefits of trade liberalization heavily flows in an one way traffic – that to the Western countries. The issue of poverty reduction is then pushed back to the discussion on quantum of 'aid'; and the current level of US aid remains 0.1 per cent of Gross National Produce or GNP! All the hypes in International Forum end up in a pittance. And the developing countries continue to face the double tragedy of inequality and poverty.

3.0 POPULATION AND POVERTY

Increase in population is often blamed for increasing poverty in a developing economy. Poor people, large in number, have nowhere to go for survival except being increasingly dependent on nature and free natural resources. 'If the total number of poor goes up, but, because of population growth, the percentage of the poor in the total population goes down, has poverty gone up or down?' asked Kanbur (op.cit.). Statistics show that even in an increase in population by more than 100 million, as seen in South Asia, incidence of poverty declined by only five per cent. Percentage reduction therefore may not be the only true indicator of reduction of poverty, weightage on the total number should be an equally important task.

The population and poverty linkage often looks like a vicious cycle; overexploitation of resources by the poor may provide opportunity to survive but once the finite resource gets depleted at faster rate, the poverty is apt to increase. To fight the poverty, increase in income becomes a logical objective and the poor often tend to increase the number of children with expectation of higher family income, with each pair of new hands, even long before they attain adulthood (banning child labour in such an economy has its own ethical conflicts). The elitist concept of population control in such a scenario falls flat as can be evidenced in the Indian subcontinent. While the number of children per family in educated middle class and the rich, significantly declined between generations, the number of children in the poor families increases. Such trends of birth rate may also be partially attributed to the fact that rate of child mortality in such families is much higher and only more children could guarantee a security to the parents when they become old. Unless social security for the poor is guaranteed (one such attempt however partial is Indian government guarantee for 100 days of employment to rural poor), the family control programmes amongst the poor will continue to fail.

Obviously per capita availability of free but finite natural resource declined with higher population count. And again poor are the worst victim of such shortage of resources. Developmental demands in India and elsewhere have almost always displaced the rural poor even from their small land holding. Jawaharlal Nehru, the first Prime Minister of India mentioned that for national interest some people have to sacrifice. And almost always poor people fighting for survival were forced to sacrifice more. Unless the population is stabilized, the pressure on ecosystems and natural resources are apt to increase undoing economic gain.

4.0 BACK TO ECONOMICS

The proponents of the MDG state 'for the first time in history, global economic prosperity, brought on by continuing scientific and technological progress and self reinforcing accumulation of wealth, has placed the world within reach of eliminating extreme poverty altogether'. That the idea is not fanciful, it is said, is evidenced in the dramatic economic progress made by China, India and low income parts of Asia' (Sachs, 2005).

'Faith' in economic progress is not unquestionable. Unless income is distributed evenly, the poor may benefit little. Real growth is therefore not a simple free market phenomenon. Cutting back spending, by the governments in low income countries to make room for the private sector may not be the miracle mantra; investment in critical areas to make the benefit reach the poor may be a more pragmatic way out.

UN millennium development goals identified the need for investment to help impoverished region to cover basic wants in health, education, water, sanitation, food production and roads for communication etc. It is estimated that requirement of assistance across the globe will be $ 160 billion a year (UN, 2005), double the current aid-assistance of $ 80 billion. To meet the goal, 0.7 per cent of GNP is to be committed, not in paper but in practice. So far it has never been done! (currently US international aid is around 0.2 per cent of GNP!). But will aid-assistance only be able to halve the percentage of poor by 2015? Unless subsidies for agriculture in developed countries is reduced (as promised) and export restrictions from developing countries as also tariff barrier is lowered, the goal may remain beyond reach.

By the time 2015 arrives, will the AIDS epidemic stop spreading over poorer nations? Can malaria be eliminated as has been achieved in case of small pox? If not economists will have to temper the assessment of poverty reduction with an account of 'missing poor people, whose deaths contributed perversely, to a reduction in poverty' (Parfit, 1984).

Lastly, a look can be taken at the current perception about development and sustainable development. UN Development Reports basically use three Human Development Indiees income, life expectancy and literacy. But Sustainable Development Indices run to more than 100 criteria, mostly non-economic, but indicative of quality of life. The question remains wide open as to which set is more reflective of a development that ensures optimum resource utilization for the present generation and assure resources for the future and really indicates the level of removal of inequality.

5.0 GLOBAL THREAT

Studies have shown that weak states with million of poor people create conditions for multiple conflicts and trigger trans-border tension. Poverty prevents economically weak countries to allocate appropriate resources to health, education, rural development or nature conservation. World Health Organization (WHO) stated that only 11 per cent of budget allocation is made in countries where 90 per cent of world's disease burden rests. Increasing civil conflicts are now being correlated with low income per capita. And poverty may ignite violence and provide ideal breeding ground to spread terrorism. As such it cannot be any longer regarded as a national concern of the poor people. USA is increasingly inclined to believe that conflict ridden Sudan and Afghanistan were selected by Al-Qaida to establish training camps (Rice, 2006).

Coming to the more near location, political history will reveal that naxalite movement had its origin in the areas of extreme deprivation; poverty and inequality was for real in the belts of North Bengal, Bihar, Andhra Pradesh; the method adopted by the leaders of Naxalite movement may be debated but the hope to achieve equality through social justice was aroused amongst the poor. The same sense of social injustice and poverty crossed the border across Nepal and at the end of blood spattered movement, democracy replaced absolute power of monarchy. Such movement in India or in Nepal didn't breed international terrorism out of poverty, as is feared by US Government for expanding terrorism reaching US shores. Tracing back 9/11 incidence at New York, American diplomatic circle became almost conclusive that cross border activities due to poverty, specially in the middle-east and elsewhere can pose global threat. It is believed that 'conflict zones not only cost lives, but they can incubate every type of trans-national security threat'. American intelligence agency CIA's Task Force opined that states in which 'human suffering is rampant (as measured by high infant mortality) are 2.3 times more likely to fail than others'.

Not that only poverty triggers unrest and violence. Hundreds and thousands of young educated people from middle class families may lend their voices and support violent movement against racials, religious discrimination. All suicide bombers are not coming from poor families but its their belief and commitment which triggers such action.

Nevertheless, Poverty remains a global threat halting the march of development towards growth and prosperity of the Human Society. As soon as the global leaders from affluent countries – the powerful G8 or European Union realize the intensity, perhaps it can help to achieve the MDG; (see India's position in the interim report given below); on either way the commitment to reduce poverty will help the developed nation in the long run, provided they believe both in cause and action. The developing countries on their part must relook at their investment in human welfare; India's investment in Defense is nearly double its investment in Health, Education and Rural development but together. Investment in public welfare at the right time in the right place may be one of the positive strategy for removal of the poverty.

Box: MDG: Country Score: India

***Extreme Poverty* – MISS:** Asia-Pacific on the whole is on track, but India, with a 2004 poverty rate of 34 % and home to a third of the world's poor, is making slow progress. More than one-third of India's 1.1 billion lives on less than a dollar a day when only one person out of 10 in China is poor as opposed to one in three in 1990.

***Child Mortality* – MISS:** Countries like Laos, Indonesia, Mongolia and the Maldives are on track to reduce their overall infant death rates. But in India, close to half of the under-five children are still underweight and 1.9 million of them are dying from malnutrition every year. Forty per cent of all underweight children in India live in Maharashtra, Orissa, Madhya Pradesh, Uttar Pradesh and Bihar.

***Disease Burden* – HIT:** India shines as the only early achiever in curbing both HIV and TB death rates. While India's 2006 prevalence was 0.36 % – over 2 million of the 6 million AIDS sufferers in the Asia-Pacific live in India – Iran and Russia have made no progress. All nine North and Central Asian countries seem to be losing their battle against TB.

***Environment* – MIXED:** India's forest cover and protected areas have increased, but it has not been able to reduce its carbon emissions or chloro-fluorocarbon (CFC) consumption. Only North and Central Asian countries have made significant progress.

***Water and Sanitation* – MIXED:** India has managed to provide drinking water to 83 % of its rural population. But it continues to house a third of Asia-Pacific's population without access to basic sanitation. Although up from 3% to 22%, the sanitation coverage is still not enough to meet the 2015 target.

***Education* – HIT:** Another success with 95% primary school enrollment. The report notes that India has helped push the region's average enrollment rate to 94%. But the report also points out that the grade 5 is low in the region – 73% in India and 79% in Nepal.

***Gender Parity* – MIXED:** While India is on track on gender equality in primary and secondary schools, progress is slow in higher education. Brunei, Philippines, Thailand, Iran and Russia have already achieved gender parity at all levels. Also, women's share of paid non-agricultural employment is generally law across South Asia – 18% in India 15% in Nepal and only 10% in Pakistan.

Table 2.1

Parameters	India	Pakistan	China	Bangladesh	Sri Lanka
$1 Poverty	■	▼	●	▼	▼
Underweight Children	■	■	●	■	N.A.
Primary Enrolment	●	■	N.A.	●	N.A.
Reaching grade 5	▲	N.A.	N.A.	■	N.A.
Primary completion rate	▲	N.A.	N.A.	▼	N.A.
Gender primary	▲	■	●	●	●
Gender secondary	▲	■	●	●	●
Gender tertiary	■	▲	▲	■	N.A.
Under – 5 mortality	■	■	●	▲	●
Infant mortality	■	■	▲	■	●
HIV prevalence	●	▲	▲	▲	▲
TB prevalence rate	●	●	●	●	●
TB death rate	●	●	●	●	●
Forest cover	●	▼	●	▼	▼
Protected area	●	●	●	●	●
CO2 emissions	▼	▼	▼	▼	▼
ODP CFC consumption	▼	●	●	▼	●
Water urban	●	●	▼	▼	●
Water rural	●	●	■	■	▲
Sanitation urban	■	●	■	▼	●
Sanitation Rural	■	▲	■	■	●
LEGEND ■ Slow, ▲ On Track, ▼ No Progress / Regressing, ● Early Achiever					

REFERENCES

Beatty, Jack, 2002. Do as we say, Not as we do. *The Atlantic Monthly*, February 2002, p. 24.

Dollar, David and Kraay, Aart, 2002. Growth is good for Poor. *Journal of Economic Growth*, 7 (3): 195.

Kanbur, Ravi, 2005. Growth, Inequality and Poverty. *J. Int. Affairs*, 58 (2): 223-323.

Parfit, Derek, 1984. Reasons and Persons. Oxford University Press.

Rice, Susan, E., 2006. The Threat of Global Poverty. *National Interest.*

Sachs, Jeffrey, D., 2005. Can extreme Poverty be eliminated? *Scientific American*, 223(3): 56-65.

United Nations, 2005. A Practical Plan to achieve Millennium Development Goals. UNMillennium Project. *www.unmillennium project.org*

CHAPTER III

ENVIRONMENT AND HUMAN RIGHTS

1.0. INTRODUCTION

Economic Development in the present day world shifted with a new approach as modern technology evolved since the days of Industrial Revolution in Europe. Capital, Labour and Technology led to an expanding market cutting across continents; the soaring profit line was good enough incentive to invest more, produce more and market more products. Need based life style changed into an era of consumerism. "Sky is the limit" concept gained ground till the decade of 1970's. Retrospection was deemed essential by a group of intellectuals; a study supported by 'Club of Rome' was published as a futuristic projection called "Limits to Growth" (Meadows et al. 1972). The 'World Model' includes equations in which the 'population growth increases with food output and in which the rates of resource depletion and pollution levels rise with the level of industrial output'. The time frame used was 1900 to 2100. It was observed that over the early path of the period, population and industrial output per capita continued to grow but model predicts that Industrial output and food per capita will plunge and at the end the standard of living is likely to sink below the level prevailing in 1900.

This was indeed a 'Doomsday' scenario. One has to remember that the American author Racheal Carson predicted a 'Silent Spring' about five years before the publication of 'Limits to Growth'. These studies had stirred the Western World. Right to Environment seems to have grown out of consciousness of human society with the observance of the 'Earth Day' in April, 1970. A new beginning was made. A re-look at the very process if developmental economics, environmental policy and quality of life, were imperative. (Baumol & Bates, 1979).

The lesson learnt become clearer with increasing awareness about the inherent problem the path of development is likely to face. While population explosion in the developing world was blamed for all the ills, the trend of consumerism in the Western World was equally criticized in view of the fast resource depletion to meet the privileged 25 percent at the cost of suffering 75 percent.

The "World Conservation Strategy" (1980) called for a paradigm shift in the methods of development. Sustainable Development becomes the key word with enormous support from the Report of the Bruntlund Commission.

2.0. LESSON FROM 20TH CENTURY

The enormous development of science and technology empowered the Human Society to achieve new dimensions of economic prosperity. The advent of space science, remote sensing, nuclear power and finally the magic of microchips made the planet earth a wonderland. In the process, the finite resources started depleting at the fastest pace – the entire biosphere became arena for this new "mantra" of development. Land, Water, Atmosphere, Forest and Vegetation, Freshwater and Marine resource – each started pathological syndrome of degradation and suffered from man made pollution load.

In May 1984, the World Resources Institute organized a meet of 75 experts from 20 countries to look at global resource and population problem and opportunities. The experts analyzed the situation and proposed initiatives for private and public action 'allaying the restive pessimism that stands between the world we have and the one we want'(C Speth, 1985). The key word that emerged is again 'Sustainable Development'. The Global Possible Conference declared that several critical transitions that must be implemented. These include

- A demographic transition to a stable world population of low birth and death rate.
- An energy transition to high efficiency in production and increasing reliance on renewable energy.

- A resource transition to reliance on nature's 'income' without depletion of its 'capital'.
- An economic transition to Sustainable Development and on broader sharing of its benefits.
- A political transition to a global negotiation keeping in tune the North and South, East and West.

It was suggested that such a prosperous and sustainable future is attainable.

3.0. TREND OF GLOBALIZATION

However, while the global consensus on the sustainable development might have emerged by end of 1990's, the drive towards more freedom of trade and breaking down barriers for facilitating 'global market' economy, was seen only when the world accepted WTO prescription. To severe monopoly interest, the system of "Trade Related Intellectual Property Rights" (TRIPS) became a buzz word. Structural readjustment and free trade witnessed a changing scenario in International Economy.

The Gross Domestic Product (GDP) in the World's fastest growing economies of China and India soared. In a boom time for generating more capital gain, the number of poor didn't decline. 'Agenda 21' adopted at the UN conference on Environment and Development (UNCED), remained an unfulfilled pledge. The chasm between rich and the poor widened.

During this process, the state monopoly over natural resource at first became much stronger but with globalization, the so called corporate efficiency in managing natural resources gained ground. Groundwater is being openly depleted by the MNC's at the cost of village water supply to manufacture "soft drinks" and "bottled water". Poor and marginalized lost their homestead land and agricultural land however small the holding may be, in the name of creating Special Economic Zone or in other developmental projects. Since Independence, at least 30 million rural Indians were evicted for only "Multipurpose River Valley Projects". Millions lost land to other development project currently being pursued by private capital with State's support.

4.0. CONSTITUTIONAL RIGHTS: POLICY

Indian Constitution has guaranteed "Right to Life" and "Right to clean Environment" but also underlined the duties of the citizens in exercising such rights. Its not only a part of fundamental rights guaranteed under constitution of the Republic, it's also the policy of elected Governments in a democracy surviving over last six decades since Independence.

The Policy statements on Environment however came much late in 1992 following action in setting up autonomous Pollution Control Boards at Central and the state level, way back in 1974. The environmental acts and rules promulgated since 1970's promise a clean and safe environment and lead one to dream a better quality of life. Acts and rules also led to setting up Standards. Incentives for clean environment and disincentives for violation of laws were clearly publicized. But in reality, offenders often managed to go scot-free. Implementation of excellent set of legal instruments remained tardy. Citizens in India continued to suffer the deprivation from clean environment.

The failure of administrative bureaucracy perhaps sensitized the civil society. Environment Movement became the new events of Indian scenario. References to such movements can be found in several publications (Krishna, 1996; Dembowski; 2001).

The issue of Human Rights over environmental resource was first widely felt during "Chipko Movement" in Garhwal Himalayas in 1970's. The women's right to collect firewood from forests was denied leading to a non- violent movement; hugging the tree to prevent clear felling by contractors, saved not only forests but established women's rights to collect dried woods, lops and tops as only source of fuel. The Government of India enacted Forest (Conservation) Act in 1980 recognizing need for re-looking at forest's role in ecosystem functioning. The women's group also is related to have organized the largest voluntary afforestation programme through eco-development camps. (Ghosh, 1989). While affected people established their rights over natural resources, elsewhere the citizens at large voiced their concern about possible negative impacts on one of the richest rain forest in Kerala due to a proposed Hydel Power Project. The years of 1980-1983 witnessed "Save Silent Valley" movement; the movement forced the Central

Government to review the project and accept people's verdict. 1980's also witnessed the now famous movement to prevent ecological disaster in the Doon Valley against limestone mining. Judicial verdict in 1985 finally enclosed 27 mining operation and led to reclamation of degraded land and afforestations. Such movements to establish people's rights can be chronicled from Koel-karo project area in Bihar through Balco project in Orissa to the most famous movement named 'Narmada Bachao Andolan', cutting across Madhya Pradesh, Maharashtra and Gujrat. While national media became increasingly sensitized over such mass based movements, state level case studies also demanded serious attention.

One such being the public interest litigation filed against conversion of 800 hectares of wetlands in East Calcutta. A movement organized by civil society and led by PUBLIC – a new born NGO, created media sensation and public interest; the verdict of the Calcutta High Court clearly went in favor of citizen's demand. Strangely, the same state government has now declared it as a "Ramsar Site"- Wetlands of international importance.

5.0. CHANGING POLICY: PARTICIPATORY MANAGEMENT

The growing concern for environment gradually made inroads towards new policy formulation. The State and the Central Government felt the urge for decentralized system of governance and need for People's participation in conservation of natural resources.

The formation of Forest Protection Committee (FPC) under Joint Forest Management (JFM) policy of 1988 and similar opportunities extended for participation in Watershed Management or Minor Irrigation Management, bear testimony to People's power. The legal sanction for a compulsory Public Hearing during Environmental Impact Study, further established people's power. The Government has at last published a 'Rehabilitation and Resettlement' Policy in 2005, a new 'Water Policy' advocated for a policy change towards 'Rainwater Harvesting'.

The 73rd and 74th Constitutional amendment empowered the local communities to establish a system of local self governance including management of the vital natural resources like Land and Water. The Panchayati Raj system advocated implementation of such a process of decentralization. The environment movement in India led the judiciary to establish special "Green Branches" in the High Courts of Kolkata and Chennai. The Supreme Courts of India also forced the government to take *suo-moto* actions, one such being cleaning up Delhi's automobile pollution through change over to CNG from conventional fuel. It is again, at the behest of a PIL, the Supreme Court forced the West Bengal Government to reallocate 400 odd tannery units in Kolkata with provision for a common effluent treatment plant and thereby minimize the Chromium pollution of surface water.

The awareness in the civil society backed up by media had enforced a positive change of attitude in the system of governance. That felling of trees and conversation of Wetlands are anti-environment, just to cite two examples, have now become a matter of of common concern.

The civii society apart, the role of Judiciary helped to establish human rights over environment. Collectively, the actions perhaps made the system of governance more alert and accountable.

The human society especially in the tropics depends heavily on biodiversity and biomass. Wild and domesticated biological resources provide all the food, fodder, a large percentage of rural fuel supply, building material, textiles, paper and pulp, medicinal and hundreds of other essentials, raw or value added such resources are now valued in billions of U.S.$. The Biological Diversity Act,2002 and Rules 2004, rightfully acknowledge the rights of the local community by suggesting Biodiversity Management Committee (BMC) at the local level; Preparation of " Peoples' Biodiversity Registers" (PBR) as a mandatory process of managing local resource is a vindication of Human Rights over local living resource.

To sum up, human society, now considered being composed of Biosphere People and Ecosystem People- the former outside the purview of specific ecosystem and the latter intimately linked with system – have been able to establish Human Rights over Environment. Considering, Environmental Awareness, even globally, has a history of only four decades, establishing Human Rights over Environment can be perceived as one of the fastest phenomenon in the modern times.

REFERENCES

Baumol, William J and Oates, Wallace, E.1979. Economics, Environmental Policy and the Quality of Life. Prentice Hall Inc., N.J.

Dembowski, Hans, 2001. Taking the State to Court: Public interest litigation and public sphere in metropolitan in India. Oxford University Press, New Delhi.

Ghosh, Ashish 1989. Environmental Movements –A Brief Review. *Sci Tech Focus* (3&4):30-32.

Krishna, Sumi 1996. Environmental Politics: Peoples' Lives and Developmental Choices. SagePublications, New Delhi, London.

Meadows, Donella et. al. 1972. The Limits to Growth: A report of the Club of Romes'. Project on the Predicament of Mankind. A Potomac Associates Book, published by Universe Books.

Repetto, Robert(ed). 1985. The Global Possible: Resources, Development and the New Century. Affiliated East West Press Pvt. Ltd.

CHAPTER IV

Right to Information: Environment and Quality of Life

1.0 INTRODUCTION

Preparation to any analytical report in India invariably faces uncertainties due to lack of baseline data. Even when data are available, often they are not accessible. One such glaring example may be cited from the first hand experience of this author (1989 – 90) during preparation of a master plan for Environment Management for one of the oldest coal field areas viz., Ranigunge Coal Field (RCF) in West Bengal. Covering an area of 1650 sq. km RCF has a long history of mining; the area looks devastated due to abandoned coal mines in absence of any reclamation measures. In this scenario, when a query was made to the "Coal Mines Health Board" on the occurrence of TB and other respiratory diseases, the answer was totally negative. No information was available according to the authorities who are supposed to record and keep data on health of the mine workers, besides other informations for more than 100 years!!

Similar experience came to our researchers working on collection of data for preparation of a "Status of Environment Report in West Bengal" in 1998 – 99. Although this project was commissioned by the Department of Environment, Government of West Bengal, many of the concerned stakeholder departments of the state repeatedly delayed / declined or provided data of little use. Similar experiences must have been recorded by researchers, concerned citizens or media representative across the country. The veil of secrecy, without any apparent justification, shrouded the realm of information of public interest.

2.0 RIGHT TO INFORMATION ACT 2005: GENESIS

The long awaited "Right to Information Act" (RIA) of 2005, therefore was expected to promise a paradigm shift from the past attitude of bureaucracy. It recognized that in a democracy, all information held by government ultimately belongs to people. Henceforth, the law promises that making information available to all concerned citizens is a part of normal government functioning. In the ultimate analysis, the people of India have a right to know 'what officials do with their money and in their name'. Access to information held by public bodies becomes a part of citizens' rights. It contradicts Official Secret Act of 1923 and brings the fresh air of openness, trust and understanding.

The demand for access to information, it can now be recalled, was persistent in 1990's. The National Campaign for People's Right to Information (NCPRI) and other such movement from civil society, made the government see a viable, implementable law come into effect. The Freedom of Information Act, 2002, remained a non – starter till the National Advisory Council (NAC) of the government of United Programme Alliance (UPA) ensured that the promise of a "progressive participatory, meaningful" right to information is fulfilled. The Right to Information Bill, 2004 was passed by the Indian Parliament on 11 – 12 May, 2005, and Right to Information Act, received President's consent on 15th June, 2005. The RTI Act finally became operational in October 2005.

3.0 THE ACT AND ITS APPLICATION

The act in the preamble states 'whereas the democracy requires an informed citizenry and transparency which are vital to its functioning and also to contain corruption and to hold Governments and their instrumentalities accountable to be governed' the act is being made 'to provide for furnishing certain information to citizens who desire to have it'. The act has 6 chapters and 31 sections; the chapters are divided into

- Preliminary
- Right to Information and obligations of Public Authorities

- The Central Information Commission
- The State Information commission
- Powers and Functions of Information Commissions, appeal and penalties
- Miscellaneous

In all the cases, the information Commissioners following First Schedule are to take oath of affirmations to discharge their duties faithfully with no fear of favour, affection or ill – will, as per section 13 (3) and 16 (3) of the Act. The second schedule of the Act under section 24 enlists the security organization established by the Central Government, who are exempted from the provisions of the Act. The same clause is applicable to all security organizations established by the State Governments. But every notification regarding exemption at the state level shall be laid by the State Legislature.

Logically, the information to be provided by the Government may not be given free of cost. Application fee imposed so far however may vary from State to State between Rs.10 to Rs.50; additional fees are to be paid for each page containing information or cost of maps or for inspection of reports (on hourly basis). Some states like Chattisgarh made differential fee for requests received from people living below poverty line (BPL) and those from above BPL.

The law has provision for appointment of Public Information Officer in all administrative units or offices at Central, State or Local Self Government levels. PIO's are to receive and process applications seeking information received from the citizens. Names of PIO's have to be prominently displayed on notice boards. At sub divisional level, Assistant PIO's are to be appointed. A single authority with multiple functions e.g. Kolkata Metropolitan Development Authority (KMDA) or Delhi Development Authority (DDA) should enable the applicants to file application with one PIO and with one time fee.

The RTI Act empowers citizens to:

- Inspect records of works
- Certified copies of documents and records

- Certified samples of materials
- Get information in Electronic form

But it does not allow to access information, disclosure of which may harm national security or information forbidden by court to be disclosed, or information which may cause a breach of privilege of Parliament or States Legislature, or information containing Trade Secrets, information on client / doctor or lawyer, cabinet paper, diplomatic correspondence etc.

The most rewarding contribution of RTI is the directive given to the authorities to proactively publish regularly, information of wide public interest. Section 4 stipulates publication of 17 categories of information on a regular basis. The RTI act opens up opportunity of the public participation in making policy – decision, (e.g. for land acquisition for development project or policy for poverty alleviation).

While the act is a powerful democratic tool, its proper application depends on following the current procedural formalities. To identify the right public authority that can provide information, and also the right person, are deemed crucial but drafting the application in an articulate way is also equally vital. Applicant must know the correct procedure to make appeal (in case of rejection) to the Appellate Authority. One has to be clear regarding difference between appeals and complain, the latter can directly made to relevant Information Commission in case of delay, excessive charges or bad decision.

The act has for the first time offered the citizens to access informations, including those on environmental aspects. The right to life has recently been interpreted by Calcutta High Court as "Right to Pollution Free Life" (January 28, 2008) and as such information on the level of pollution – air, water, noise may be of vital interest to any citizens. But pollution apart, can this act be used to elicit information on such environmental resources like wetlands in Urban area or Parks, Gardens and open space in Cities and Town? The act ensures access to information already generated through public funding by Government Departments – but what about the information that the concerned authority do not have a clear record or data base?

The data deficiency becomes more apparent if one attempts to analyze such data – set. In a recent study on provisions of essential environmental services in 123 Urban Local Bodies (ULB) in West Bengal, it became apparent that many of the ULB never provided data to the empowered Government Institution i.e. Institute of Local Government and Urban Studies (ILGUS); as such these remains vital data gap on percentage of households with piped water supply, sanitation facility and drainage & sewerage connection. Obviously no worthwhile analysis can be carried out of in such context (Ghosh, A. K. 2008). Such non availability of data, however, can make the citizens empowered with the new Act, and make the concerned bodies work specially when public funding was spent to provide such services. A Delhi based NGO 'Parivartan' used the Delhi Right to Information Act 2001 to probe the extent of utilization and accountability of Public Works; it was revealed that items worth Rs.70 lakhs were missing out of investment of Rs.1.3 crore spent for 68 Public Works. (Devasher, 2006).

Environmental issues concerning Health and Nutrition, Occupational Health, Renewable Energy, Groundwater resource etc. can all be investigated with reference to the investment of public money and actual performance of the agency vested with the responsibility. Informed citizens can only question the system of governance; otherwise evasive answers will elude the civil society.

One such possible query may be to test the actual application of extant law e.g. Water (Prevention & control of pollution) Act 1974 and Air (Prevention and Contrl of pollution) Act 1981. In the both the acts, there are clear provisions of fiscal penalty and imprisonment for violation of the provisions of the act. Till date, no one really knows how many people have been convicted, penalized or imprisoned? Obviously such disclosure helps to deter the violators of laws.

The effectiveness of the Act can only be judged from the Annual Report that Central Information Commission or State Information Commission is to prepare (u/s 25). Such report will indicate the number of requests made to each public authority, the number of decisions on denial to access, the numbers of appeals referred to authorities, particulars of any disciplinary action taken against any officer for lapse in administration of the Act, the amount of charges collected and recommendations to reform or amend the Act.

It is also to be noted that Government, within financial and other limitation, should develop and organize educational programme, especially for disadvantaged people as how to exercise rights contemplated under the Act. So not only enacting the vital law, responsibility of disseminating the contents to the citizens lies with the Government.

Unfortunately, not many evidences of such effort is visible in most parts of India. In a resent meeting on the role of media in publicizing RTI Act (1st March 2008) held at Calcutta it was pointed out that the concerned government agencies as well as the visual and print media across the country have failed miserably to make the people aware of the provisions of the RTI Act. Questions can be raised on the logic of spending public money, advertising every single occasion of foundation-stone laying ceremony, as against absence of any campaign to make the people aware of RTI provisions. Any sample survey would sure to revel that common people still remain ignorant about the provisions of the Act. There is hardly any publicity on the part of Information *e*Broadcasting Ministry/DAVP to make it known to the people. Even the print media remains rather shy to focus the power given to the People of India.

Unless the provisions of the Right to Information Act 2005 is made widely known to the citizens through major Indian language media, AIR, Newspapers and Television Channels, the citizens can hardly make use of the same. After 73rd and 74th amendment, the system of governance has been decentralized, but the decentralized bodies are not geared to take up the new challenges to make the citizens empowered to ask and receive information governing their life and livelihood.

The 'Quality of Life' can only change with an alert citizens group. Access to information is the power that civil society can very well utilize and make use even in Public Interest Litigation (PIL) challenging wrong deeds of the system of governance be it at Local, State and Central Government level. Even if the government fails to make RTI Act widely known, the civil society organization access the country can take the onus of making it known to more than one billion people of India. That such efforts are now being made can be witnessed form some recent publications. (Bhushan, 2006, Gadgil, 2006, Raina, 2006) Unless steps are taken through concerted efforts very limited application and 'Business as Usual' scenario will prevail.

REFERENCE

Bhushan, Prashant, 2006. Right to Information Emasculated. *Combat Law,* 5 (5): 20 – 21

Devasher, M, 2006. Your Guide to Using the Right to Information Act 2005. Commonwealth Human Rights Initiative, New Delhi. 1 – 85

Gadgil, Madhab, 2006. Science and the Right to Information. *Economic and Political Weekly,* 61(19): 1895 - 1902

Ghosh, A. K. (ed) 2008. Status of Environment in West Bengal – A Citizen's Report. ENDEV Society for Environment and Development, Kolkata.

Raina, Vinod, 2006. Killing the Bill. *Seminar*, 565: 29 - 32.

CHAPTER V

LAND RESOURCES: NEED FOR A POLICY

1.0 INTRODUCTION

The conservation of natural resources has been a major focal area in Country's economic development. The land resources act as the backbone of agriculture, forests, industries and sources of biomass. The total geographical area of India, standing at 328.73 million hectares (mha) being finite, appropriate utilization of land and soil should be a major thrust area for resource development in the country.

The pressure of human and livestock population has led to degradation of 173 mha of land out of total geographical area of 329 mha. Per capita cultivable land has been reduced from 0.32 ha., in 1951-1952 to 0.17 mha., in ?-1992. Annual loss of nutrients is estimated at 8.4 m to 15.37 m.tonnes (Bali, 1997).

Table – 5.1: Land Degradation in India (million hectares)

Geographical area	329.00
Area subject to water and wind erosion	144.43
Area degraded by special problems	•
Waterlogged area	8.530
Alkali soils	3.158
Saline soil including coastal sandy area	5.150
Raverines and gullies	3.197
Shifting cultivation	4.910

Riverine and torrents	2.730
Total problem area	173.65
Annual loss of nutrients	15.37 to 8.4 mtonnes
Total Drought-prone area	60.00
Total Flood-prone area	40.00
Average annual area affected by flood	9.0

[Source: M. S. Rama Mohan Rao, 1994 and D. L. Dab, 1994]

Agriculture being the dominating land use the area when calculated comes to nearly 190 mha out of the reporting area of 304.89 mha., representing 62% of the land.

Of the total land area, reporting area is pegged at 304.89 mha. Forests occupy nearly 77 mha., of land area being variably distributed over 31 states and seven Union Territories, the highest forest area being located in Madhya Pradesh (now divided between Chattishgarh and Madhya Pradesh), followed by Andhra Pradesh, Arunachal, Maharastra, Orissa and Uttar Pradesh (now divided between Uttaranchal and Uttar Pradesh) – collectively contributing nearly 51 mha. Normally forest cover denotes an apparently healthy land scenario and as such only seven states of India could be listed as in possession of good land resources (67%); however, a more critical analysis will reveal the extent of degraded forest land in the country. Forest area occupies 23% of the land in India.

Area put to non-agricultural uses accounts for 6.8% while barren and uncultivated land represent 5.8%. The percentage of land used for different sectors however do not represent the actual profile of land and soil.

2.0 PROBLEM AREAS

Vohra (1980) mentioned, 'category wise 61% of our agricultural land and at least 72% of our non-agricultural lands are degraded to a greater or lesser degree'. It was mentioned that three quarters of total relevant land area is in need of urgent attention and a third is so sick that it is almost unproductive. India's Environmental Action Plan (EAP, 1993) recognized as one of the major problems confronting the country is land degradation affecting 174 mha comprising agricultural, nonagricultural and forest land. Between 1980-1993, the Wasteland Development Board was established in 1985 and EAP (op.cit) reported that 40 mha of degraded land including the land affected by water logging (8.5 mha), ravin and gallies (4 mha), saline and sodic condition (9 mha). Under the strategies part, EAP recommended programmes for rehabilitation and catchment area treatment for major and medium irrigation works and formulation of water management plan based on detailed soil surveys and land use capabilities for major and medium irrigation projects. It can be noted that linkage between land degradation and faulty irrigation programmes of the country was the single emphasizing factor in EAP.

That irrigation projects often undermined prevailing soil characteristics is now widely acknowledged. In soils, which lack natural drainage characteristics, 'the presence of excessive surface water results in a rise of level of subsoil water, till it reaches the root zones of crops'. The resultant effects are loss of productivity and turning the land totally barren.

The contribution of development projects by engineering works e.g., embankment for flood control, road and rail and canal embankment. Often such projects do not pay heed to the necessity of cross-drainage works.

In order to increase agricultural production, canal irrigation has become a major thrust area. Often canal irrigation projects lack appropriate field channels and adequate provision for surface and subsurface drainage. EAP's focus on command area development. That leveling and re-shaping of command area may be an integral part of irrigation project is now well understood but scant attention is being given to do the same.

Besides the problem of water erosion, the issue of wind-erosion also calls for serious attention. Overexploitation of non-agricultural land by felling green cover on overgrazing. When such land belongs to the State Government or Common Property Resource (CPR), it becomes more vulnerable; in absence of grass or vegetation, the soil become exposed to erosion by wind leading to loss of top soil. An earlier estimate shows that area under wind erosion account more nearly 16% of land under reporting. Creation of windbreaks and shelterbelts can reduce velocity of wind and prevent erosive and desiccating effects. Field experiments in arid zone of Rajasthan shows that once the pasteurs can be re-established, it can support your times more animals but only through controlled and rotational grazing.

Till 1980, conversion of forests for non forestry purposes was an uncontrolled activity. While the Forest Conservation Act, 1980 demands prior approval for such conversion, a slowing down process between 1980-2000 seems to have reversed; an unprecedented rate of conversion of forest to meet demand of development mainly for multipurpose river valley projects and specially for mineral exploitation. Simultaneously, inspite of participatory forest management introduced since 1988, a large scale degradation of forest land has been recorded all over India. The loss of canopy cover led to expansion of open forest area to 8.8% as against very dense forest figure of 1.6% (FSI, 2003).

Mining activities have increased largely due to expansion of Coal and Iron ore mining in Jharkhand, Chattishgarh, Orissa but other mining of minerals like Bauxite, Limestone, Dolomite have also expanded over 15 years (1990-2005) to meet increasing demand of Cement and Aluminium industry. Inspite of the fact that more environmental friendly technologies being available for underground mining (e.g., long-wall mining in Coal mines) and open cast mines with mineral deposit under shallow layer (e.g. Bauxite), wide spread adoption of such technology has never happened. The end result is vast tract of wastelands left by the old mines. Currently environmental planning is necessary for new or expansion of mining area lease but real test is with the process of mine closure and reclamation plan. The degradation of land and soil due to mining is not confined only to the mining area *per se* but it involves transportation of machinery during construction and mined material to supply point during operation. Often such process necessitates removal of dense forest cover. While construction phase activities can not be avoided, transportation of mined material by conveyor belt / ropeway to designated load points may prevent such deforestation process and consequent loss of top soil.

The open cast mining accounts for major mineral output. In the coal sector, open cast mining activities increased by nearly 20% between 1981-82 and 1999-2000. While new open cast mine need to prepare reclamation plan, monitoring of execution remain far from being complete. Once top soil and subsoil is lost, land reclamation of open cast mines becomes very difficult (Khoshoo, 1986). Besides the environmental cost of soil degradation, loss of agricultural production (as in the case of Asansole – Ranigunj Coal Field in West Bengal, Ghosh, 1997) and loss of forest cover, if added may provide a different profile of cost benefit analysis. Since mining can never be stopped, adoption of appropriate planning and technology and post mining monitoring becomes essential.

3.0 STRATEGY FOR RECLAMATION AND CONSERVATION

The Government led programme to reverse soil degradation has achieved some success but could never accomplish the target; as a matter of fact the targeted figure has often been revised to make it more realistic or cover facture. Late Prime Minister set up The National Wasteland Development Board in 1985 and assigned it to afforest and reclaim 5 mha., of wasteland annually; the figure had to be revised to 1.5 mha. In 1992, Prime Minister Narashima Rao set up Department of Wasteland Development but participatory management remained a far cry. However the state of Madhya Pradesh has achieved a path breaking success in Jhabua district. Intelligently using watershed management 2.9 mha of land was being made green (CSE, 1999). But such examples are still far and few to see. People's initiative has changed wasteland in Puruliya district of West Bengal, in the degraded forest of Himachal and some other parts. But it still remains a basically government controlled programme for reclamation and conservation.

The National afforestation and Ecodevelopment Board (NAEB) was set up in 1992 specifically to address increasing loss of forest cover. The board has made a draft plan to achieve target of 33% forest cover by 2012, including Multi Stakeholder Partnership (MSP) framework for afforestation of degraded forest land with private capital.

By separating the task of recovering degraded forest land from the other Non Forest area Wasteland Development, precious little has reportedly been achieved. The decision of the Government of India to allow private participation to grow trees in degraded forestland, in the long run will not much help the nation but assist the paper and pulp industry to procure raw material; such a process will lead to loss of rural income from the farm forestry.

An institutional credit system for development of India's wasteland was proposed by NABARD based on an Action Plan for 15 years. The author of such a proposal M.S. Haque, Deputy General Manager (Tech) based his calculation with a target of reclaiming 75 mha., of wasteland which needs immediate attention; the master plan was supposed to develop 5 mha., of wastelands every year; the calculated cost was Rs. 15,000/- per ha. One can remember the original target of reclamation set up by Prime Minister Rajib Gandhi; it was scaled down not because of lack of funds but lack of initiatives.

Institutional rearrangement may not be the key word for soil conservation and wasteland development. India is having more than 4000 Civil Society Organizations concerned with environment. Participatory wasteland development programme is however yet to take place at a national level. Indian Council for Forestry Research and Education (ICFRE) has spent significant amount of research fund to offer mining wasteland reclamation technology; but it has never been actually field tested over wide area. Technological options for Wasteland Development, which includes water-eroded lands, wind-eroded land, ravines and gullies, salt affected soil, waterlogged land, acid soil etc., are all supposedly available. The reported fund requirement was grossly estimated at Rs. 15,000 crores in 1990's. But the actual implementation seems to be a distant dream. There is more data coming up on degradation, causes and monitoring system but the shadow of loss of land and soil seems to be ever increasing.

4.0 DISCUSSION

The spurt of national policies published by the Government of India – nearing 20 since 1991, however does not include any policy on Land and Soil. The stakeholders of this vital resource base, especially the primary users at the grass root level continue to suffer because of lopsided attitude of the governments ruling the country under different political system.

The problems of combating desertification, reclamation of wasteland, soil erosion and consequent silt load in the surface water system on one side and cost of environmental damage on the other, needs a comprehensive understanding. Creation of multiple organizations under various ministries, enactment of stringent laws and rules without a monitoring mechanism has so far failed to reduce the problem at the national level. Research data generated by multiple agencies and academicia remain without resultant effects.

The Panchayati Raj Institutions may be the right platform to launch a programme with a time target but then again political motivation and willingness of the elected representative at the village level needs a great thrust.

In 1992, the National Conservation Strategy and Policy Statement on Environment and Development was announced by the Ministry of Environment and Forests, Government of India. In the said policy 'conservation of land and water' was specially emphasized under the section 'Conservation of Natural Resources'. 'Enactment of laws for appropriate land uses to protect soil from erosion, pollution and degradation' was considered essential.

In 2000, the National Agriculture Policy also emphasized on 'abiding importance in improving the quality of the country's land and soil resources. It has elaborated on the need for reclamation of degraded and fallow lands and problem-soils, management of land resources on watershed basis, management of pasteur land etc.

But in reality, the situation in 2006 remains problem ridden. Country needs a real National Programme for Renewal of Land and Soil Resources, with adequate funding provision, technology adoption and accountability of concerned state governments.

REFERENCES

Bali, J. S., 1997. Wasteland Development and Watershed Management – Concepts, Importance and Need. p. 1-11. In Advances in Wastelands Development. Soil Conservation Society of India, New Delhi. 242 pp.

CSE, 1999. State of India's Environment-5. The Citizens Fifth Report. Centre for Science and Environment, Part-I, 440 pp.

FSI, 2003. State of Forest Report, 2003. Forest Survey of India, Ministry of Environment and Forests, Dehradun, 134 pp.

Ghosh, A. K., 1997. Degradation of Land in Rajigunj Coal Field. p. 133-139. *In* Advances in Wastelands Development. Soil Conservation Society of India, New Delhi. 242 pp.

GoI, 2000. National Agriculture Policy. Government of India, New Delhi, 9 pp.

Khoshoo, T. N., 1986. Environmental Priorities in India and Sustainable Development. Indian Science Congress Association, New Delhi, 224 pp.

MoEF, 1992. National Conservation Strategy and Policy Statement on Environment and Development. Government of India, New Delhi, 36 pp.

MoEF, 1993. Environmental Action Programme, India. Ministry of Environment and Forests, Government of India, New Delhi, 220 pp.

Vohra, B. B., 1980. A Policy for Land and Water. Department of Environment.Government of India. Sardar Patel Memorial Lectures, 1980, 28 pp.

CHAPTER VI

LAND RESOURCE: POVERTY, INEQUALITY AND RIGHTS

1.0 THE FINITE RESOURCE: INFINITE DEMAND

Land is one of the finite resources, the human society depends on. Within a given geographical area and political boundary of a nation, land resource within state control has faced severe management problems. Obviously, the demand and supply market gets highly strife ridden in absence of a policy on land resource. Land rights have become an increasingly focused issue in India, as well as the need for land reforms.

India with nearly 329 million hector lands has in reality little over 300 mha, for which land use data at micro level is reported. Agriculture and Forest still strongly dominate the use and ownership of India's land. Land reforms were introduced in India in the decade of 1950's. While national guidelines were laid down, land reforms were made a state subject. It has subsequently been seen that the 'land reforms measures adopted by the Indian Union or respective states are not in accordance with any personal law of any community' (Gupta, 2002). Even in West Bengal, when land reforms have surpassed that of any other states in India, women remain second in the land agenda.

The affirmative management of India's land can only be vested with the people. Mahatma Gandhi said 'political Independence is only the first step towards true independence. The second step would be to give ownerships of the land to those who are tilling it'. Being predominantly dependent on agrarian economy, where agriculture continues to be the prime source of livelihood for a majority, land is taken as the 'pivotal property in terms of both income and employment.

The abolition of Zamindari system in 1950 was perhaps the first step towards securing land rights for people. Zamindars in many areas refused to submit rent rolls and other relevant land records; the law was amended in 1954 and 1959. Before 1950, there were 20,5,907 permanently settled holdings representing 90% of the land area. Even after abolition act, Zamindars reportedly were successful to keep at least 14% of the land besides occupying large tract of land as *khas* land. The abolition of Zamindari left agrarian laborers mostly Dalits and the sub tenants to the lurch.

2.0 WHO OWNS LAND?

Land ownership therefore provides livelihood options and economic security in countries like India. Landlessness is equated with deprivation poverty and vulnerability. The so called third world countries, including India, witnessed violent peasant movement due to inequitable and unjust distribution of land. A recent statistics show that while 45 percent of global population still depend on land for livelihood, about 500 million people remain absolutely landless. Majority of the 1.2 billion poor people in today's world live in rural landscape but without access to land.

Indian independence evoked hope and aspiration to millions of poor. In contradiction to the wishes of the Father of the Nation that 'Peasants would take the land they cultivate and no compensation would be paid to landlord' more than Rs.6000 million was paid compensation to the landlords. (Louise, 2006)

The hope of millions was belied. The Promised Land reforms, considered revolutionary, remain an unfulfilled dream in vast rural scenario of India.

An assessment shows that since 1950, a total of 73,51,431 acres were declared surplus of this 7.3 million acres, nearly 87.5 percent was taken possession but less than 70 percent of the land taken possession, was actually distributed. The lack of political will resulted in non-implementation of policy. Till 1994, only 4.9 million people have benefited from redistribution.

Performance of State Governments however vary widely. The figure denoting as surplus land is also said to have been manipulated politically. Large landlords still exist in North India; in Uttar Pradesh, of the 5,11,028 acres that has been taken possession of, only 3,68,174 acres have been distributed; in Himachal Pradesh, only 3340 acres have been distributed out of 2,81,462 acres taken over from surplus land. A look at the All India figures (Table 6.1) can reveal the situation about a decade back but not much has changed since then.

It's not the frustrating profile of land reforms in the country that makes one concerned, it's also the changing ownership of land in the name of economic development that makes us think. India being the 3rd largest dam builder in the world, has evicted more than 30 million people during last 60 years from their homestead and agricultural land to make way for River Valley Projects. Considering that the country lacked a proper Resettlement and Rehabilitation Policy till 2005, the plight of these helpless, faceless millions of citizens can easily be guessed.

The denial of access to surplus land on one side and deprivation from legally possessed land by force of antiquated law, made millions of rural Indian a sad witness to a mega serial of 'Development'. If poverty alleviation is a national agenda, the biggest driving force that could have made it happen is land reforms. But all the talks of land reforms appear rhetoric in reality.

Reclamation of degraded land is another unfinished agenda. The 'technology – mission' mode of wasteland reclamation launched by Late Prime Minister Rajiv Gandhi in late 1980's has suffered repeatedly due to changing target, lack of actual supervision and willingness of the participating states. So the problem of land-reforms is compounded with the problem of landless and land degradation.

3.0 LAND RIGHTS: A NEW MOVEMENT?

The denial of access to land has perhaps the worst impact on the Dalits and the Tribals. Till five years back (during 9th five year plan period) it was stated that 90 percent tribals and 77 percent dalits are fully or partly landless. This is despite of as many as 270 legislations passed over land reforms since independence. It is shockingly revealing that during last five decades only 1.25% land in India was redistributed as against 43% in China, during last 50 years. Sharma (2006) aptly mentioned 'obviously, when 7.43% big

Table 6.1 : Land Reform Implementation

States	Area Declared Surplus	Area Taken Possession of	Area Distributed to Individuals	No. of Beneficiaries	S.C. Beneficiaries Area in Acres	S.C. Beneficiaries No. of Beneficiaries	S.T. Beneficiaries Area in Acres	S.T. Beneficiaries No. of Beneficiaries
1	2	3	4	5	6	7	8	9
Andhra	800968	571632	510989	437176	218781	189614	101543	77662
Assam	612380	575837	482868	424364	44138	42364	57389	40964
Bihar	474621	399932	278719	330101	162482	199255	38128	40631
Gujarat	2480320	156003	130335	29814	80552	13007	28685	12871
Haryana	121409	115932	113124	38018	59738	17657	0	0
Himachal Pradesh	284053	281462	3340	4400	2305	29334	139	261
Jammu & Kashmir	455575	450000	450000	450000	0	0	0	0
Karnataka	273878	159716	117153	72084	70346	54012	3002	1573
Kerala	136363	91330	64078	143763	25108	61530	5152	7353
Madhya Pradesh	288572	256580	185673	72209	49752	21601	73516	26991

1	2	3	4	5	6	7	8	9
Maharashtra	723103	648366	556806	139457	160236	41781	101049	28351
Manipur	1830	1685	1682	1258	128	96	97	72
Orissa	174007	132573	151242	130880	48184	44956	63362	48695
Punjab	133876	105153	102406	26634	41801	10072	—	—
Rajasthan	609646	552423	440624	76762	139057	28246	44174	10737
Tamil Nadu	184972	171114	149650	124834	59607	56435	233	178
Tripura	1995	1944	1598	1424	212	256	448	359
Uttar Pradesh	542618	511028	368174	30120	250581	217932	1332	745
West Bengal	127068	1201684	941382	2059576	349630	764836	180775	395429
D & N Haveli	9406	9305	6811	3313	48	25	6760	3286
Delhi	1132	394	394	654	277	495	0	0
Pondicherry	2355	1207	1022	1359	612	797	0	0
Total	**735143**	**6428393**	**5058070**	**4887750**	**1753580**	**1758984**	**705788**	**696158**

[Source: National Commission for SC and ST. Annual Report 1993 – 94]

farmers own 41.57% of land, compared to 31.80% land cultivated by 64.33% small and marginal farmers, then only Five Year Plan doesn't serve the purposes and priorities of poverty reduction in India'.

The professed promises of land reforms programme can be recalled as follows:

- Abolition of intermediaries.
- Regulation of rent.
- Security of Tenure.
- Ceiling on land holding.

Redistribution of land, it is evidenced in many developing countries, to the landless or near landless farmers can actually help in reduction of poverty, increase food production at low cost and empower marginalized community. Emphasizing role of small farms in increasing food production, FAO of United Nations, urged, in its publication (Agriculture towards 2000) to release only 5% of the farmland in India; this land with improved access to water, could reduce rural poverty by 3%. Poverty alleviation in India can be linked with land and water reforms in the country.

While the government system continues to fail, people's movement gained momentum; NGO's are categorically demanding access to land for the poor and landless. Demands include:

- Deadline for preparation of revenue village wise roster of all pending cases.
- Setting up land Dispute Settlement Tribunal, composed of officials and nominated stakeholders.
- Arrange for Public hearing.
- Forming a State Level Coordination Committee of Members in charge of Revenue, Land Reforms, Law and Justice, Social Welfare, Home and Panchyati Raj.

Civil society organizations, like 'Ekta Parishad' (EP) were born in Madhya Pradesh in 1980's. The organisation has now spread to eight states across India from Bihar, Orissa, Jharkhand and Chattisgarh in the east to Uttar Pradesh in the north, Gujarat in the west, on to Tamil Nadu and Kerala in the south. Following Gandhi's philosophy of 'Satyagraha', the organisation through massive campaign, covering 3500 km., and 10000 villages within six months (December 1999 – June 2000), could make Madhya Pradesh Government from a 'Task Force' for facilitating land redistribution among tribals and dalits; as many as 3,47,000 marginalized families have reportedly benefited from such a process. Similar positive action based programme benefited 500 *Baigas* families in Chattisgarh. Since the first "Padayatra", EP has organized similar marches in Bihar, Orissa, Jharkhand and Uttar Pradesh. Similar campaigns have also been organized by "Research Foundation for Science Technology and Ecology" New Delhi and many other NGO's to establish land-rights.

4.0 THEN AND NOW

Historically, demand for land rights started from Gandhian 'Bhoodan Movement' led by Binoba Bhave in 1950's. Jayprakash Narayan (JP) opined that 'Bhoodan' is not a charity but is to establish the demand for realization of human rights. JP led movement in 1970's lent further credence to Peoples voice. One has to remember the earlier movement of 'Tebhaga Andolan' in 1950's as a part of demanding rights over the produces from the land, launched by the Marxists in West Bengal. The movement saw Police brutality with aim of suppressing the voice of protest against injustice.

But situation has since changed in many parts of India. For instance, in West Bengal, it is claimed that "the land reform programme was definitely the most radical that India has witnessed in the recent period". (Anon, 2004)

In West Bengal, Government had undertaken a major change in tenant relation involving entire "recording or registration of sharecroppers who had cultivated on the same piece of land for a number of years. This registration with the Department of Land Revenue gave them permanent and inheritable rights to cultivate the land". The legal permission had already existed, but never implemented in the past, till 1970's. It is claimed that land reforms played an "important role in unleashing productive forces in cultivation and improving the balance of class relations in the countryside in favour of the poor and working classes and peasantry generally' (ibid, p.42). But at the end it is admitted that the economic process that generate inequality in the countryside still remain significant. NSS (National Sample Survey) data indicate a sharp rise in the number of landless rural people, from 39.6% in 1997-98 to 49.8% in 1999-2000. Nearly half of the rural households still remain landless in the most lauded state of West Bengal (as against national average of 41%)! It is believed that West Bengal scenario is due to substantial diversification of rural employment to non-agricultural sector.

What ever may be the reason, the fact remains that in India, poor are yet to get the justified share of land – a resource which could alleviate poverty. Recent controversy over Tribal Rights on Forest land, where they have lived for generations is yet another pointer. The Scheduled Tribes (Recognition of Land Rights) Bill 2005, was opposed by deep ecologists with the apprehension that it will lead to destruction of the forest. (Soni, 2005). But historically, forest villages settled by British Rules were for handy labour; now forestry operation has changed and villagers do not have jobs or earnings. The advantages of Revenue villages are denied; they remain in eternal poverty. This apart, the association of tribal's with the forest is like umbilical cord between child and the mother. Establishing legal rights will enable the community to pursue productive work. Tribals will never kill the forests which provide their life – sustaining system. Traditionally they have protected the forests and with rights established they will do so again. Seven percent of the Indian population is Tribal – how, for decades their rights have been denied?

The other raging controversy is about land acquisition in the name of development, without going into the details, the fact that this antiquated law is now going to be revised is a welcome feature. The question – is it fair to ask farmers to sell land even if they don't want to. This however is interlinked with fair price and compensation in the form of Rehabilitation and Resettlement. The R&R Policy is yet to be concretized and a new act is on its way in 2008. Public opinions have been invited in a recent media advertisement (See Box: item 1, 2, 3 out of 7). It has been emphasized that the new step Rehabilitation and Resettlement Bill, 2007 is an attempt to give a statutory backing to the provisions of The National Rehabilitation and Resettlement Policy, 2007. The said legislation proposes to provide for the basic minimum that all the grievance of the affected persons. The Bill also contains a saving clause to enable the State Governments, Public Sector Undertaking or Agencies, or other requiring bodies to continue to provide or put in place greater benefit levels than those prescribed under the Bill. Expectedly policy of forceful eviction in a democracy in the name of development will change and more human face will emerge.

BOX

PARLIAMENT OF INDIA

Feb 11, 2008

LOK SABHA SECRETARIAT

STANDING COMMITTEE ON RURAL DEVELOPMENT (2007 – 2008)

SUGGESTION ON THE LAND ACQUISITION (AMENDMENT) BILL, 2007 AND 'THE REHABILITATION AND RESETTLEMENT BILL, 2007'

'The Land Acquisition (Amendment) Bill, 2007' and 'The Rehabilitation and Resettlement Bill, 2007' as introduced on Lok Sabha have been referred by Hon'ble Speaker, Lok Sabha on 7 December, 2007 to Parliamentary Standing Committee on Rural Development headed by Sri. Kalyan Singh, Member of Parliament, Lok Sabha for examination and report to the Parliament. The text of the Bill as introduced in Lok Sabha is available at website **'http://loksabha.nic.in'** under the icon **'New Events'**2. The Land Acquisition (Amendment) Bill, 2007 seeks to amend the various provisions of The Land Acquisition Act, 1894 with a view to strike a balance between the need for land for development and other purposes and protecting the interests of the persons whose lands are statutorily acquired.3. Besides, The Rehabilitation and Resettlement Bill, 2007 is an attempt to give a statutory backing to the provisions of the 'The National Rehabilitation and Resettlement Policy, 2007. The said legislation propose to provide for the basic minimum that all the grievance of the affected persons. The Bill also contains a saving clause to enable the State Governments, Public Sector Undertaking or Agencies, or other requiring bodies to continue to provide or put in place greater benefit levels than those prescribed under the Bill.

While the land reform policy continues to suffer from non-implementation, the process of globalization ushered in the era of Special Economic Zone (SEZ). The SEZ Policy of India offers benefits to the developers and export units a set of lucrative incentives viz.,

- Duty free import/domestic procurement of goods,
- O & M of SEZ units,
- 100% Income Tax exemption on export income for first five years, 50% for next 5 years and 50% of the ploughed back export profit for the next 5 years,
- Exemption from minimum alternate tax,
- Exemption from Central Sales Tax, Service Tax and State Sales Tax and other levies,
- Tax exemption for external commercial borrowing upto $500 million in a year etc.

But many fears and apprehension have been raised. A major fear is caused by large land acquisition in the name of development. Agricultural farmers have become most vulnerable. Big corporate sector may simply relocate units to escape taxation offence (website till February 2008 announce 154 notified SEZ occupying 20,388 hectors or 204 sq. km's). Rao (2008) former Director General of National Council of Applied Economic Research commented 'In some, their development (i.e. SEZ) might in fact be a major land grabbing effort by organized industry together with the government'. The government has so far been unable to publish full details about all approved SEZ's. Little information is available on monitoring and regulation of SEZ's. But land resource is going off to large corporate ventures.

In conclusion, absence of a firm land resource management policy and denial of rights over such vital resources has been creating chaos and confusion, while abject poverty continues to belie Millennium Development Goals of the United Nations. One has to remember that, India, has pledged to go by MDG.

REFERENCES

Anon, 2004. West Bengal Human Development Report. Department of Planning and Development, Government of West Bengal, 202pp.

Gupta, Jayati, 2002. Women Second in the Land Agenda. *Economic and Political Weekly,* 37 (18): 1146 - 1154.

Louise, Prakash, 2006. Land Rights Discourse: A Critique, *empower.org,* 2(6): 3 - 5.

Rao, S.L. 2008. Special Blindness: there is little hope that SEZ in India will be of much use. *The Telegraphs,* Calcutta, February 11. 2008 p. 8.

Sharma, Ramesh, 2006. Land First: Peoples Participation in land rights. *empower.org* 2(6): 6 - 7.

Soni, Vikram 2005. Tribal People and Preserving Prime Forests. *The Hindu,* Friday, September 29, 2005, p. 12.

CHAPTER VII

WATER CRISIS

1.0 INCREASING DEMANDS

Water resource around the world is increasingly facing an unparallel crisis; crisis of management, crisis in sharing, crisis in supply to meet ever increasing demand. Water resources have uneven geographical distribution. The UN World Water Development Report of March 2003 predicted that up to 7 billion people in 60 countries, more than the current global population will face water crisis by 2050. The report revealed that demand of water has tripled over past 50 years and the trend is going to be in an ascending curve to cope up with population and accelerated economic activities in a globalized world.

In India, the demand sector is dominated by the Irrigation need – 85 per cent of water is said to be spent in agriculture, globally the figure is about 70 per cent. Finding the major reasons for water crisis, as usual, will come back to the consumerist human society. The consumer based society is apt to use more water even when not needed; such a practice has its own impact in the total water scenario. Wastewater carrying enormous pollution load only help to reduce the useable resource level. Large scale conversion of wetlands into dry land became a critical issue for concern. Wastage from inefficient supply line contributes significantly, specially in the developing countries. Add to this the impending crisis of climate change which is likely to 'account for 20 per cent of the increase the global water scarcity' – says an UN Report.

In the following parts an analytical framework is proposed based on selected issues viz., watershed management, dams for development, climate change etc.

2.0 MANAGEMENT CRISIS

Managing urban and rural water supply have now been termed an area that deserves better understanding. Mismanagement of say urban watershed has been analyzed critically in some countries e.g., USA (Platt, 2006). Decentralized experiments in collaborative management is now being strongly advocated. Complete watershed partnership is being advocated for stream restoration. Major goals however can be categorized into Water Quality improvement, Flood hazard reduction, Aquatic habitat of Fisheries restoration and Public use and awareness. A series of strategies both 'Site Specific' and 'Non Site Specific' have been suggested (Table - 7.1).

Table – 7.1

Urban watershed Management Goals and Strategies

Goals	Strategies	
	Non-Site Specific	Site Specific
Water quality improvement	Non-point source controls: Septic system upgrade SSO Control / sewer repair Maintain minimum flows Low-impact development Litter clean up Monitoring	Point source controls: NPDES discharge permits CSO reduction (LTCRs) Green stormwater BMPs Riparian bioengineering Brownfield remediation Reforestation / aquifer recharge
Flood hazard reduction	Flood modeling and mapping Floodplain regulations Warning / evacuation plans Flood proofing Automated rainfall / streamflow monitoring Public information	Remove channel obstacles Restore natural flow regime Flood prone property buyout Wetland restoration Green stormwater BMPs (such as rain gardens, green roofs, and porous paving)

Aquatic habitat / fisheries restoration	Maintain minimum flows Increase dissolved oxygen Reduce toxics, organics, metals etc. Reduce bank erosion and sedimentation Litter clean-up Invasive control (such as the zebra mussel) Fish restocking	Dam removal Stream daylighting Remove shoreline armoring Restore stream flow regime (pool / riffle/ meanders) Riparian buffer restoration (bioengineering and reforestation) Invasive plant removal / planting of native species Wetlands restoration Fish passageways on dams
Public use and awareness	Watershed public events Reduce health hazards Improve visibility of stream Public information (such as signage, websites, and newsletters)	Urban waterfront renewal Green ways / bike ways Public recreation sites Boat launch ramps Environmental education sites

[Source: Platt, 2006]

A critical look at the suggested strategies and comparing Indian experience may be of interest. Non Point Source Centres in India over last 33 years (since the Water Act of 1974 has been passed) will reveal a frustrating set of data. While septic system remains non-existent in even many small urban settlements, sewer repair issue does not come. The existing sewerage system in large and old urban metropolises (e.g., Kolkata) has been upgraded in sectoral manner under different time frame but overall results provide no solace. Domestic wastewater still provides the bulk of pollutants to the urban stream.

Lack of flood plain regulation and warning evacuation plan, has been witnessed in massive Mumbai flood in early 21st Century. No automated rainfall / stream flow monitoring helped to control urban flood. Known aquatic habitats, specially the large water bodies have no inflow control system as witnessed in Kolleru Lake, Andhra or Lake Chilika in Orissa. All the funding spent for Public use and awareness during much hyped "Ganga Action Plan" had little impacts; success in case studies like Benaras city in Uttar

Pradesh, India is attributed to individuals rather than to institutional efforts. So all the strategies suggested by Platt (op.cit.) may not yield the desired results unless the system of governance become effective. The list of strategies, site-specific ones, also demands strong and determined actions by the designated authority, which appears to be half hearted or sadly lacking in initiative. The success of National River Conservation Plan (NRCP) proposed by the Ministry of Environment and Forests will invariably depend not only on strategic planning but on determined action. The formation of National River Conservation Directorate under the Ministry of Environment and Forests to implement River and Lake Action Plans was a welcome move but the objective of improved water quality, upto bathing class standard, of 34 major rivers of India calls for critical questioning. The experience of Ganga Action Plan (GAP) may be acting as a bitter reminder.

The pollution abatement works in 160 towns spread over 20 states of the country at a cost of Rs. 4735.42 crore, will be keenly watched during ensuing 11th Five Year Plan period.

Urban watershed management is not the only area of crisis. Uneven distribution of water across the country is naturally based on surface flowing streams and rivers. Following models of mega projects in western countries, specially Tenesse Valley Authority (TVA) in USA, India installed the grandeur project of Damodar Valley Corporation (DVC). The objectives were to control flood, regulate water flow and provide water for irrigation and also to generate pollution free hydel power. In a recent conference held at Department of Business Management (February 23, 2007) at the University of Calcutta, US, Fulbrighter Danielle Klingesmith pointed out that DVC and TVA were inspired by Technocrats without community consultation or participation. The long term impacts when assessed against professed objectives will show the range of adverse impacts. Reservoir depths for instance declined due to silt load, thereby changing the projected capacity and schedule of periodic release. In both the project, TVA and DVC, hydel power generation declined and coal based Thermal Power Plant (TPP) have been commissioned. In TVA, 1,25,000 people were displaced as against 1,50,000 people in DVC but while in the former evictees were paid adequate R & R package, victims of DVC specially tribals still lack the very essential items like Water, Healthcare and Education; a visit to Panchet hill area will verify the observation.

Its not only the history of DVC and TVA but the entire process of Dam building was seriously questioned in the report on 'World Commission on Dams'. But water resource experts specially in the emerging economies in China and India still heavily advocate such a strategy. Ecological process demands optimum utilization of finite resources; the trend of maximizing production of food crops defies such a process. While irrigated land increased by five folds between 1900-2000, reduction of food crop yield was significant but now attributed to salt build up in soil in many tropical countries. Canal irrigation can have serious adverse impact in water logging and increasing salinity of crop field (Ghosh, 2007). The Mega Dam mania is best exemplified in Narmada Project – with 30 large dams, 135 medium sized dams and 3000 small ones, projected flooding of 37,000 hectares of prime forest and farmland and displacing more than 1,00,000 people. Apparently there is no lesson learned exercise from the thousands of Dam building exercises carried out in the past (Readers may be referred to a publication 'Dams, Rivers and People' – a journal published by South Asia Network on Dams, Rivers and People, Delhi, *www.sandrp* in to get a periodic update on the issue).

Issues of managing watershed or Dams for development apart, a projected crisis in water management now discussed is centered around global climate change. Admitting climate change is a reality in 21st Century, a look at the possible impact of such change on Rainfall and Water Resources will be worthwhile. In case temperature rises there will be increased evaporation combined with regional changes in precipitation; these two factors in turn can potentially 'affect mean run off, frequency and intensity of floods and droughts, soil moisture and water supplies for irrigation and hydro electric power generation (Bhatt and Sharma, 2002). Add to this the real time changes in the mountain glaciers in permafrost area, denuding future potential of freshwater and simultaneously adding unpredictable volume of water to the surface flow.

Mitra and Bhattacharya (2002) estimated that runoff in the mountainous region could increase by 11% due to Green House Gas emissions (Carbon dioxide, Methane, Nitrous Oxide, Chlorofluoro Carbon etc.) over a 50 years period. Evidences from recent years indicate that chances of extreme rainfall leading to flash flood are fairly established as also the slope failures in the Himalayan region. With the shadow of uncertainty about extent of impact, the population boom continues and per capita availability of water in India has been projected to decline to 300 m^3 in 2050 from the present National average of 2664 m^3!!

The Indian Institute of Tropical Meteorology (IITM), Pune has carried out analysis of spatio-temporal variability of precipitation over the major river basins of India based on time series data. In has also estimated past changes in the surface water availability to understand their sensitivity to climate change. The study predicted that the hydrological cycle will be more intense in future with both higher annual rainfall and drought. The study also predicted increase in extreme rainfall and rainfall intensity in Ganga, Godavari, Krishna basin towards the end of 21st Century. Number of days of rainfall will show decline in the western parts of Ganga basin but increase in most parts of Godavari and Krishna basin. Such results obviously call for better planning and design of hydrological structure, a better strategy for flood control, drought control a new look at the agriculture policy for selection of crops and a better industrial and urban planning (Climate Change Impacts on Water Resources in India, Keysheet 5, Ministry of Environment and Forests, Government of India).

The management of water resources as such is not confined to the area of civil engineering of public health or irrigation departments; it calls for a holistic approach and new strategy from multi disciplinary avenues.

3.0 MANAGEMENT OPTIONS

In the preceding part only three key issues related to management crisis of water resources have been discussed. It is imperative that serious changes in the outlook and approach could not save from water crisis in next millennia. Pollution, degradation, dam building, Carbon and Methane emission are not going to change in near future. What is redeeming is the new realization emerging out of multi disciplinary approach.

Water Resources could only be sustainably utilized with planned afforestation and soil conservation measures in the catchment areas of the rivers, managing watershed with appropriate pollution control measures and going for alternatives to the 'big dam' syndrome. Micro watershed, Micro hydel are practical ideas, never grandiose but easily manageable; as such a relook at mega project and search for practical alternative will be positive task. Curbing carbon emission and cutting down GHG has already been accepted as an International agenda to reverse the current trend of climate change. India is an active partner in this endeavor.

While interlinking of rivers as an alternative to unequal distribution of water was proposed by National Water Development Authority (NWDA), serious questions have been raised but still remains unanswered (Ghosh et.al., 2003). Thankfully, concept of rainwater harvesting for community use, use at personal household level or recharging aquifer has gained ground as a part of national and state policy.

One of major issues that deserve attention is 'public consultation'. Stakeholders in water resource management are hardly taken into consideration for the decision being taken by bureaucrats and technocrats. Whenever participatory irrigation management (PIM) or management of arsenic free water for drinking has been launched, significant results have been obtained (North Bengal case study for PIM and South 24 Parganas case study for arsenic free water supply). Pani-Panchayats and other people's organizations, wherever formed are largely working well.

It may be too optimistic to visualize a paradigm shift in the mindset of present power of governance but given opportunities and marginal technical support, communities can positively suggest sustainable water use.

REFERENCES

Bhatt, J. R., and Sharma, Subodh, K., 2002. Impacts of Climate Change in India and Climate Change related Activities. P. 110-172. *In* "Climate Change and India" Issues, Concerns and Opportunities. (eds. P. R. Shukla, Subodh, K. Sharma, P. Venkata Ramana). Tata McGraw Hill Publishing Company Limited, New Delhi.

Ghosh, A. K., 2007. Environment: Issues and Concern. APH Publishing Corporation, New Delhi.

Ghosh, Asish, Sikdar, Pradip, K., and Dutta, Asoke, K., (eds) 2003. Interlinking of Indian Rivers. ACB Publications, Calcutta.

Mitra, A. P., and Bhattacharya, Sumana, 2002. Climate Change and Green House Gas Inventories: Projection,s Impacts, and Mitigation Strategies. p. 76-109. *In* "Climate Change and India" Issues,

Concerns and Opportunities. (eds. P. R. Shukla, Subodh, K. Sharma, P. Venkata Ramana). Tata McGraw Hill Publishing Company Limited, New Delhi.

MoEF, 2006. Annual Report 2005-2006. Ministry of Environment and Forests, Government of India, New Delhi.

Platt, Ruthford, H., 2006. Urban Watershed Management. *Environment*, 48 (4): 28-42.

Swain, Janet, L., 2003. Water Scarcity could overwhelm the next generation. *World Watch*, 16 (4): 4-5.

CHAPTER VIII

Biodiversity and Climate Change

1.0 INTRODUCTION

The issue of climate change and global warming has attracted attention of the scientists and statesmen alike across the world. In 1980's, the application of computing power made it possible to develop general circulation model (GCM's). These models help to understand the functioning of world weather system; it gives a global, 3-D simulation of the dynamic behavior of atmosphere based on physical properties of its constituents. The outcome of researches predicted 1.5-4.5 ^{0}C increase in temperature by end of 2000; the average warming of 0.5^{0}C predicted in the model for 20th century has now become real. Interactions of physical constituents with plants, animals and micro-organisms that strongly influence surface properties and the abundance of green house gases, is yet to be fully understood. Life on land and sea is intimately involved in climate control. The International Geosphere Biosphere Programme (IGBP) established in 1986 raised six key questions viz., (I) what is the role of biological process in producing and consuming trace gases? (II) how will global change affect terrestrial ecosystems? (III) how does vegetation interact with processes of hydrological cycle? (IV) how will land use, sea level and climate alter coastal ecosystem? (V) how do ocean biogeochemical processes influence and respond to climate change? (VI) what significant climatic and environmental changes occurred in the past and what were their causes? (IGBP,1992). For each of these a research project was initiated, outcomes of which are making significant contributions to our process of understanding.

2.0 GLOBAL BIODIVERSITY: NEW ASSESSMENT

The Convention on Biological Diversity (CBD) also attracted global attention with regard to the issues like prospecting biodiversity for future human development, access and transfer of genetic resources with prior informed consent and installing a mechanism of benefit sharing between the conserver and the provider of biodiversity and the users for commercial gains. The relationship between CBD and Framework Convention on Climate Change are now being investigated with participation of International scientific community. Exploring global change and terrestrial ecosystem exercises were focused on developing scenarios of biodiversity change for 10 terrestrial biomes and freshwater ecosystem by 2100.

Five major drivers of biodiversity change at the global scale have now been identified viz.,

- Changes in land use
- Climate
- Nitrogen deposition
- Atmospheric concentration of Carbon dioxide
- Biotic exchange (introduction of alien species)

Each one of these drivers can have significant impact, depending on sensitivity of each biome. Interactions amongst these drivers may be totally negative or antagonistic or synergistic.

Of the five drivers land use change may be considered not directly related to climate change but it is said to be the most important driver that may affect habitat availability for species with changes in forests to grassland to agricultural lands. The second important driver is Climate change; major effects of climate change on biodiversity are expected to occur due to warming at a higher latitude.

Nitrogen deposition due to human action is likely to lead to a reduction in species diversity; species with higher relative growth rate will make use of increases in nitrogen and out compete the slower growing species (Tilman, 1993).

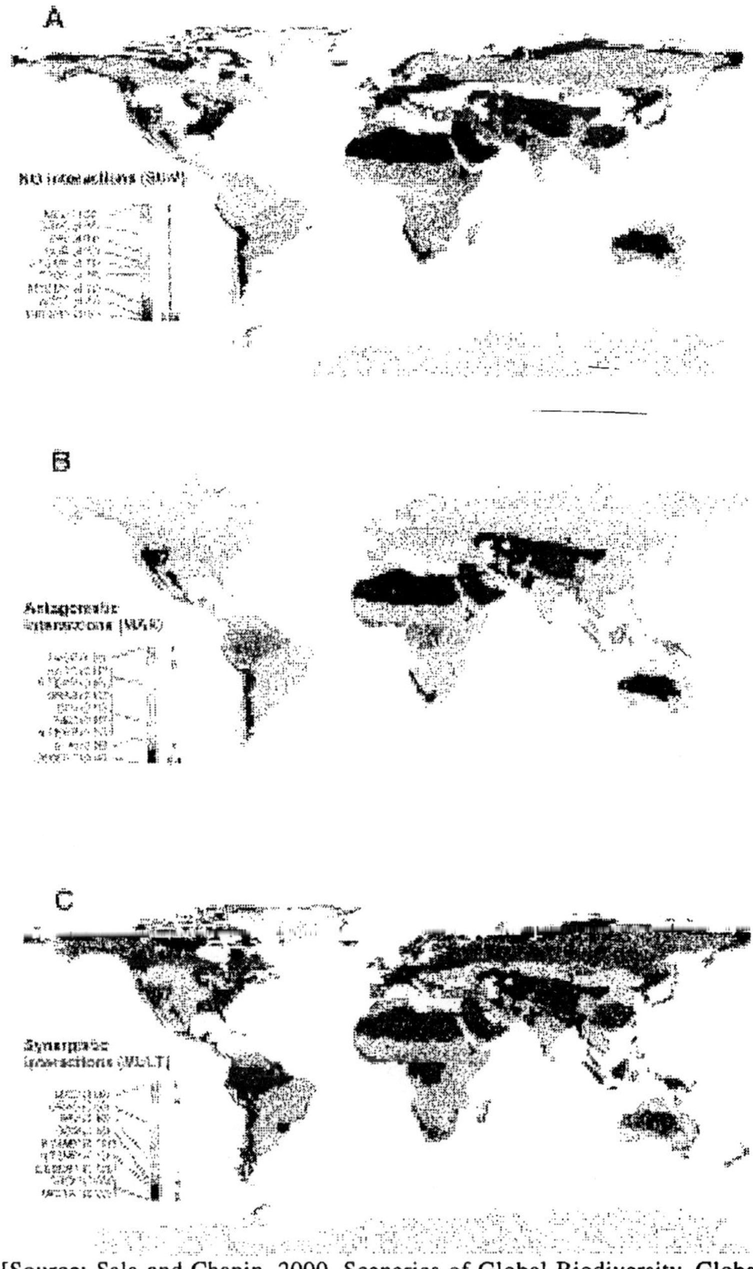

[Source: Sala and Chapin, 2000, Sceneries of Global Biodiversity, Global Change Newsletter, 43: 10]

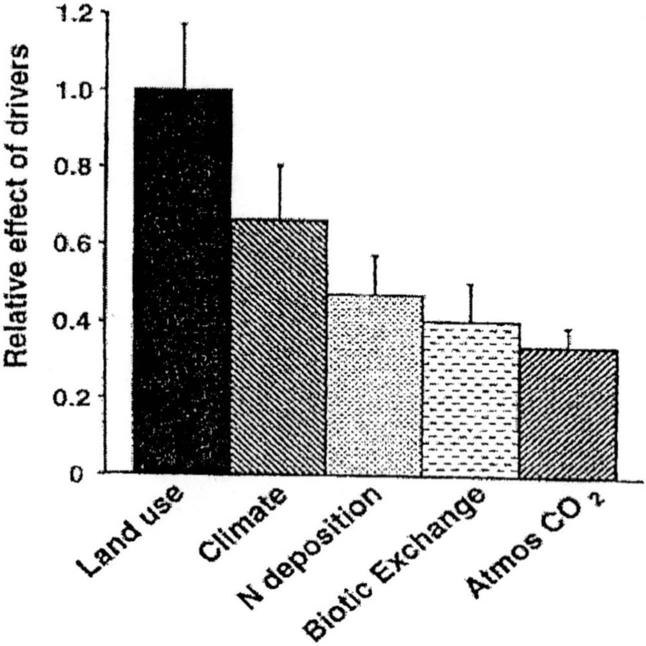

[Source: Sala and Chapin, 2000, Sceneries of Global Biodiversity, Global Change Newsletter, 43: 8]

This may lead to extinction at local or even at a global level. The increase in the level of CO_2 is likely to have effects similar to increase in nitrogen. CO_2 enrichment may help some species, which are to out compete others.

However, all these predictive models also indicate that impacts may significantly vary in different biomes (viz., Arctic, Alpine, Boreal, Grassland, Savanna, Lakes, Mediterranean, Desert, Northern Temperate, Southern Temperate, tropical and Streams). Southern Temperate and Tropical Biomes will have maximum impacts on biodiversity due to land use changes, while Arctic biome will have most severe impacts of climate change. Freshwater biome will also have considerable impact due to climate change.

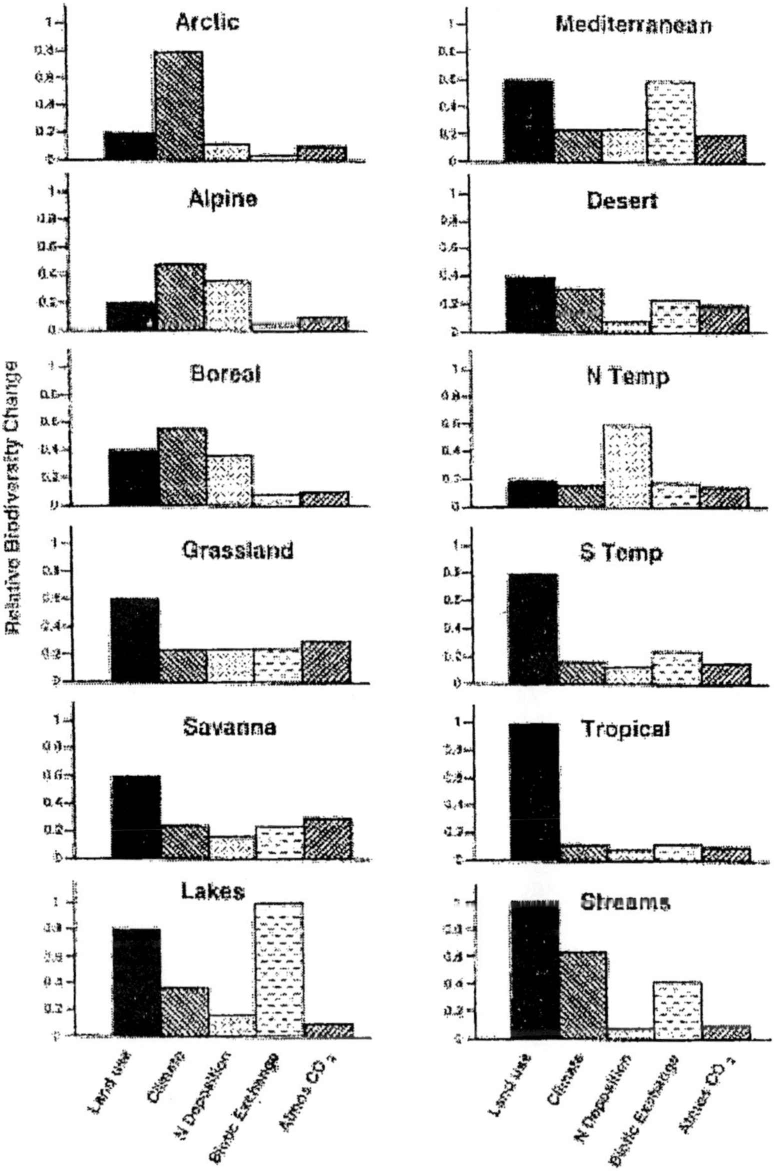

[Source: Sala and Chapin, 2000, Sceneries of Global Biodiversity, Global Change Newsletter, 43: 9]

Of the three possible scenarios, examples of synergistic interactions among drivers are many (Sala et al, 2001; Chapin et al, 2000). A larger CO_2 response is noted when N availability was high. Carbon dioxide and nitrogen deposition may have lesser impact on lakes and streams than on terrestrial ecosystem but acidic deposition (mainly due to nitrogen deposition) and its interactions with climate change, land use and stratospheric ozone depletion may be very large, specially for boreal lakes. Cumulative effects of such events may be responsible for fast depletion of global freshwater biodiversity 'at a far greater rates than is true for even the most affected terrestrial ecosystem'.

3.0 CLIMATE CHANGE AND BIODIVERSITY

Taking a cue from earlier discussion, any serious analysis will reveal the potential of exciting, multidimensional approach to investigate the issue, not withstanding the fact that non-climatic changes can also affect local, short-term biological changes. Applying diverse analyses in more than 1700 species, researchers have shown that recent 'biological trends match climate changes'. Significant range shifts averaging 6.1 km per decade towards the poles or 6 meters per decade to higher altitudes have been documented through Global meta-analyses. The studies also revealed significant mean advancement of spring by 2-3 days per decade. Migratory birds, amphibians and other animals are breeding earlier in springs and plants have early blooming season; this is evidenced in 677 species based changes in biological cycle. If global warming continues, species are apt to move southward and lower elevations during winter, just as the opposite trend may be seen during summer. Extinction of species is likely to be at a faster pace if they fail to adapt to new climate. A diagnostic finger print of temporal and spatial 'sign-switching' responses are now predicted by twentieth century climate trends (Parmesan and Yohe, 2003).

In a recent study, Stanford University biologist, Elizabeth Hadley (online edition: September 2004 of the Journal Plos Biology) reported decline in population of two mammalian species by 40-50% (Montane Voles and Pocket Gophers) during Medieval Warm Period (850-3350 AD) when their habitats dried up and an increase during the Little Ice Age (1350-1950) when the climate was wetter. These results established a direct correlation between climate change and population size. Genetic analysis shows differential behavior between two species. While Voles showed an influx of new genes and genetic diversity even when their population declined (which means they were connected to other population) Gophers showed a trend of inbreeding indicating they haven't really recovered from Medieval Warm Period. The time series data from 1000 BC to 1950 AD, covering nearly 3000 years helped to understand the possible impact of global warming. The study 'Genetic Response to Climate Change: Insights from Ancient DNA and Phylo-chronology' showed scope for future researches on genetic impacts from climate changes *(www.plosbiology.org)*.

Major research findings support (I) range-loss due to climate change, thereby jeopardizing the concept of declaring protected area (PA system), (II) faster rate of extinction, (III) change in biological cycle and migration pattern, (IV) change in genetic diversity, and (V) maximum impacts on aquatic system.

Dwelling on the possible impacts of climate change on fragile ecosystem like in the Sundarbans, Ghosh (2002) has used the predictive model of surface-ocean stratification, which may have two opposite affects of Phytoplankton for productivity. It may reduce nutrient supply from deeper layer and increase light availability due to shoaling of the upper layer, which in turn may decline primary production. By the end of 2100, the expected increase in atmospheric carbon dioxide may lead to three fold rise in surface water carbon dioxide concentration, relative to pre-industrial value; these changes are likely to affect sea water pH and enzymatic activities of phytoplankton. One has to remember the planktonic biomass provides the very basis of aquatic biodiversity including the trade in fisheries (Richesell et al, 2001).

4.0 DISCUSSION

The evidences of possible negative impacts due to climate change in biological diversity are being collected from field observations and simulated modeling. In India, the distinct bio-ecological zones are apt to react differently due to synergistic impact of the drivers of the change. Unfortunately, no systematic time series analysis is being made regarding either the flowering time of plants or animal migrations of wild animals in different habitats or altitudes. Neither impact of climate change on productivity of planktonic biomass in eco-fragile areas of

Sundarbans or in other coastal zone is being studied (at least the results of any India based study is yet to feature).

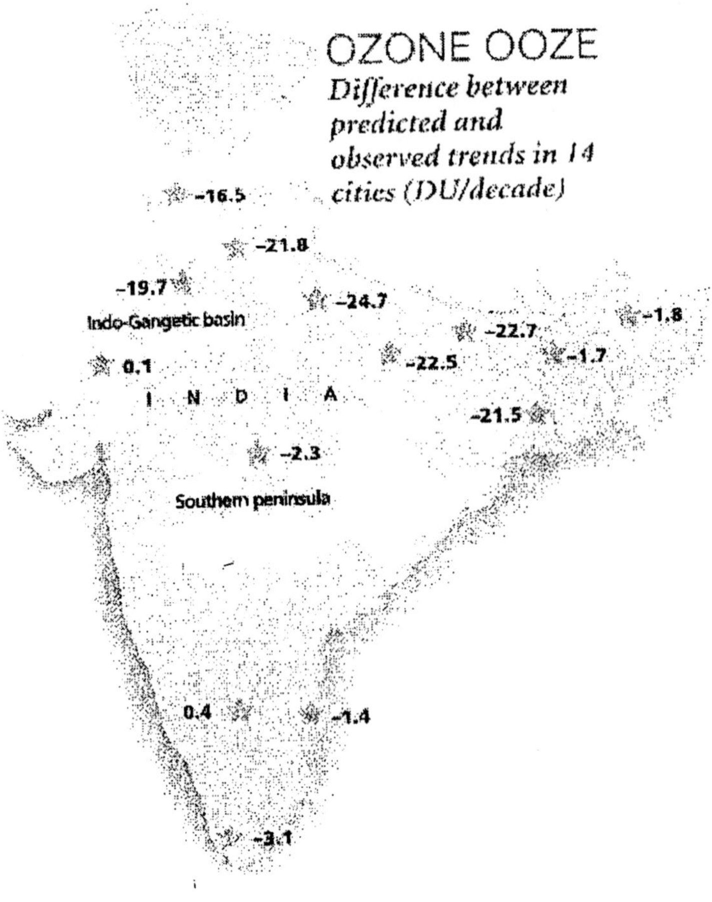

The scope of investigating impact of climate change in one of the mega-diversity countries like India is enormous. Since this involves both terrestrial and aquatic system in different biogeographic region, a coordinated project with a common approach and methodology may be the most desirable pathway. India has already witnessed definite footprint of climate change both in the coastal and peninsular region. Indo Gangetic plain has recently been identified as most vulnerable to the process of ozone depletion. A report from IIT, Kanpur and George Mason University,

USA revealed that decadal variation in the ozone column over the densely populated Ganga basin is almost three times the predicted trend for the decade ending 2003. The reason for such alarming change in ozone layer is yet to be determined; but a study report from IARI, New Delhi showed that 26 million hectares of rice-wheat cropping system in Indo Gangetic plain releases huge quantity of nitrous oxide which can destroy stratospheric ozone (*Environment Monitoring and Assessment, vol. 77, no. 2*). A media report indicates that a team of 70 scientists from 22 institutions across the country have undertaken a major study sponsored by ISRO and DST in the Ganga basin in December 2004; continuous observations of atmospheric gases, ozone, solar radiation and water vapour from eight different locations were carried out. (Aerial Threat to Ganga basin, *Down to Earth*, p. 34, April 30, 2005).

Similar collaborative work on impact of climate change on wild biodiversity is urgently needed in Indian region.

REFERENCES

Chapin, Stuart III et al, 2000. Consequences of Changing Biodiversity. *Nature*, 405: 234-242.

Ghosh, A. K., 2002. Climate change and Threat to Biodiversity in Sundarbans. 54-63. *In* National Resource Conservation and Environment Management, ABP Publishing Corporation, New Delhi.

IGBP, 1992. Global Change: Reducing Uncertainties, International Geosphere Biosphere Programme: A Study of Global Change. International Council of scientific union, 40 pp.

Parmesan, Camille and Yohe, Gary, 2003. A globally coherent fingerprint of climate change impacts across national system. *Nature* , 421: 37-42.

Riebesell, U. et al., 2001. Effects of increasing atmospheric CO_2 on phytoplankton communities and the biological carbon pump. Global Change Newsletter, 47: 12-15.

CHAPTER IX

Future Biodiversity Scenarios in India & South-East Asia

1.0. INTRODUCTION

It is recognized that 80% of recorded biodiversity- species, land races and crop cultivars are confined to the tropical countries. Asian region besides Africa and South America is well known for rich biological diversity; countries of mega diversity largely belong to these continents except for Australia. Much of the ecosystems are shared between politically divided countries who should own equal responsibilities for conservation. For instance mountain ecosystem in the Himalaya is shared between Afghanistan, Pakistan, India, China, Nepal, Bhutan or the mangrove ecosystem in eastern India shared by India and Bangladesh. Trans Boundary Rivers provide such a scenario; when freshwater system of Ganga and Brahamaputra, for instance is considered, the role of China, India, Bangladesh and Nepal become evident with reference of aquatic biodiversity in this huge freshwater resource of South East Asia. South East Asian Countries also share the same marine ecosystem with India and Pakistan in the west and India, Srilanka, Myanmar and Maldives in the South and East. All these geographical information provide the basis for biogeographical assessment of biodiversity.

The need for regional co-operation among the South East Asian Countries to achieve biodiversity conservation beyond political boundary was first focused exactly after two years of UNCED in 1992 at Rio de Janeiro. A regional meet of SAARC,

ASEAN and other countries was held at Bangalore, India (22-23 August, 1994) under the aegis of Ministry of Environment and Forests, Govt. of India and United Nations Environment Programme.

2.0. STATUS OF BIODIVERSITY

An attempt would be made to understand current status of biodiversity at ecosystem and species level, since information on genetic diversity for most of the South East Asian countries remain poor. Ghosh (1995) has presented a biodiversity analysis report on South Asia.

Bangladesh (21°N & 27°N latitude and 88°E & 93°E longitude) covers 14.4 mha of geographical area dominated by vast wetland ecosystem. The forest in Bangladesh covers 11-14% of the land area and the hill tracts in Chittagong along with mangrove forests in the Sundarban provide other aspects of ecosystem diversity. The species diversity include 125 species of Mammals, 600 species of Birds, 85 species of Reptiles, and 20 species of Amphibian besides 145 genera of commercially important fish species. No authentic data on other aquatic biota or largest group of arthropods, i.e., insects is available.

Floristically, about 5000 species of angiosperms, 500 species of Medicinal and 18 species of Bamboo, Rattan are documented. Bangladesh has no authentic national collection of its biological diversity and has alarmingly low rate of protected area (was 0.7% in 1995, might have increased under the present Government).

Bhutan (88°45'E - 92°10'E longitude and 26°45' N - 28°30' N latitude) is well known for its forests, and high level of biodiversity within the "Hotspots" of Eastern Himalaya. Sharing Palaearctic and Indo-Malayan zone, a wide range of floral and faunal diversity can be recorded. Forest types include (i) lower tropical foot hills, (ii) temperate, and (iii) sub-alpine zone. With more than 60% forest cover, it has 23% under protected area (PA) network. Bhutan shares much of its major biological resources with those of Darjeeling-Sikkim Himalayas and Nepal Himalayas in terns of mammals, birds and angiosperms. The avian fauna of Bhutan has 552 species as against entire Indian region with 1200 species.

Nepal shares the same Palaearctic and Indo-Malayan biogeographical zone. Forested area covers 37% of the total geographic area, distributed between tropical Sal forests, scrubby vegetation and alpine tree forests of Pine, Oak, Birch, Fir and Hemlock; 13.8% of land area of Nepal has been put under PA. Nepal with only 0.03% of Global space boasts of 2.1% angiosperms, 3.2% ferns, 5.3% bryophytes, 2.7% lichens, 4.3% of mammals, 8.5% birds, 1.5% reptiles, 1.1% amphibians and 2.1% freshwater fishes. The altitudal range, climatic condition and the ecosystem diversity – mountain, forests, freshwater lakes and rivers, makes this tiny Himalayan country unique.

Pakistan: Ecologically north western part of India has much in common with the neighboring country of Pakistan, sharing semi arid and arid zone. The floral and faunal diversity of Pakistan is well documented. In same groups the amphibians, Pakistan has much fewer species (27) than in India (204 species); the other group of herpeto fauna i.e. reptiles also appear poor in Pakistan (174 species) than in India (450 species). Obviously geographical area, forest range and ecosystem diversity play important role in determining species diversity. While India has a rich ecosystem diversity, Pakistan has largely been dominated by mountain, marine, freshwater and desert ecosystems with significantly less forest cover (5.2%). Species diversity includes 6000 species of plants, 188 species of mammals, 740 species of birds, 174 species of reptiles and 700 species of marine invertebrates. Pakistan's forest cover of 5.2% offers only lass than 1 mha of protected areas under 10 national parks. Protected species include 31 species of mammals, 20 species of birds and 5 species of reptiles.

Sri Lanka: With forest cover of 22.2% of land, the island nation has a history of 150 years of biodiversity documentation. Ecosystem diversity ranges from marine to freshwater, forest and mountains and grasslands. A total of 14% of land area has been kept for in-situ conservation. Srilanka has rich elements of endemics with 24 % flora, 12% mammals, 20% birds, 72% reptiles, 19% amphibian and 16% fishes being recorded as endemics but 16% of flowering plants and 28% of fern and fern allies are considered threatened, besides 30% of the fish species, 62% of amphibians, 67% of reptiles and 24% of bird species (Ghosh, 1995)

Maldives: A conglomerate of 1190 islands of which 200 are inhabited. Maldives is well known for its rich fish (900 species) and coral (± 97 species) species diversity. Information on many of the species group appear inadequate and scanty. *'Maldives'* as an island nation has been considered as one of the most vulnerable nation exposed to sea level rise due to climate change.

India: India occupying 2.4% of global species (329 mha) exhibits a rich ecosystem diversity of mountains, forests, freshwater, marine, islands, mangroves and coral reefs etc. India has been recognized as a country of "Mega biodiversity" with 8% of global biodiversity and posses two of the 25 global "Hotspots" known for uniqueness, high endemism, rich species diversity etc. India has more than 600 "*Protected areas*" in terrestrial and aquatic habitats, and 23% of area under forest cover. But land use changes and impacts of global warming are considered serious threats to India's biodiversity. India's efforts for *ex-situ* conservation through a network of more than 110 zoological and botanical gardens, gene banks and bureaus of Genetic Resources (Plants, Animals, Fish, Microbes) deserve special mention.

Malayasia: Other than the countries of South Asia, now in SAARC network, Malayasia and Thailand in South East Asia may be considered in this discussion. Since, Malayasia has a significant endowment of biodiversity, a look into its resource base and vulnerability may be taken. Malayasia with only 33million hectares is nearly 10% of India, but has a forest cover over 60% of its total land area. A network of terrestrial protected (30% of land area) area, apart Malaysia has 40 marine and 6 fishiness protected area. With 15,000 flowering plants, 300 species of mammals, 254 bird species, 400 species of reptiles and 200 species of amphibians. Malayasia has outlined strategies and plan of action for conservation of biodiversity. But currently it has listed 681 plant species and 50 mammals, 37 birds, 21 reptiles and 20 fish species as "Threatened". Ecologically, the country is dominated by forest, coastline and marine ecosystems.

It may be seen that South East Asia has three known "Hot Spots" based on species richness, endemism and other criteria (Myers 1988, 1990). These are Eastern Himalaya encompassing Nepal, Bhutan and NE India and Western Ghats. Besides SW Srilanka (Table1). These areas show a significant number of endemics amongst plants, mammals, reptiles and swallow-tailed butterflies. India is also recognized as a country of Mega diversity (Table 2). While no separate data is available for SE Asia, Asia on the whole has an impressive 13.6 area under wilderness as against South America's 20.8% and Africa's 27.5% (Table 9.3) not all wilderness areas posses rich biodiversity but countries in the tropics promises to have so . Asia has also highest number of Centers of Plant Diversity and endemic bird areas: India alone has 6 centers of plant diversity (Table 9.5 and Table 9.6).

Table 9.1 : Biodiversity Hot Spots in SE ASIA: Endemics

	Plants	Mammals	Reptiles	Swallow-tailed Butterflies
India, Nepal, Bhutan Eastern Himalayas	3500	3	20	-
IndiaWestern Ghat	1900	7	91	5
Srilanka South West Srilanka	500	4	-	2

Table 9.2 : Mega diversity Country in South East Asia

	Mammals	Birds	Reptiles	Angiosperm	Swallow-tailed Butterflies	
India	390	1230	453	17500	7	7

Table 9.3 : Distribution of Large Wilderness Area

	Km^2	Wilderness % of total
Asia	377,858	13.6
Europe	138,553	2.8
South America	3,745,971	20.8
Africa	8,232,382	27.5
North America	9,077,418	37.5

Table 9.4 : Centers of Plant Diversity & Endemic Bird Area

	Plant	Bird
Asia	83	63
Europe	21	6
South America	46	42
Africa	33	32
North America	6	14

Table 9.5 : Countries With Most Centers of Plant Diversity

	Plant	Bird
India	6	-

3.0 FUTURE BIODIVERSITY SCENARIOS

It has been stated that "Modern fortune tellers can be distributed with the 'Malthusian pessimists' at one end and 'technological optionists' at the other. Human influence on Biodiversity was a theme during preparation of Global Assessment Report and two distinct dimensions have emerged – the first being "Understanding the degree of which past human actions have set in motion irreversible" and the current changes in the physical and natural environments influencing options. The second dimension involves "the capacity of human societies to understand, adopt and respond to environmental changes" – a function of the cultural, social, economic and political contexts in which they operate.

Analyzing the trends, the question of growth in finite environment is apt to come up. And invariably the trend of future growth of one single species, *Homo sapiens* becomes the critical deciding factor. Demographers predicted way back in 1995 the global human population will double in next 60 years. The trend of population growth in South East Asia may be used as an indication to understand the shape of things to come.

The population of South East Asian selected countries shows a density varying of 49 per sq.km. to 1035 per sq.km. The mid 2007 data shows a total of 574,000,000 which is expected to reach 691,000,000 by 2025 and 778,000,000 by 2050; (Table 6.). Unfortunately, it is said that this is going to happen even if fertility rates fall in virtually every developing country (Jolly and Terrey, 1993). Such a trend will drastically alter per capita availability of all finite resources in South East Asia by 50% reduction. The human species is said to approximate 40% net primary productivity of the terrestrial ecosystem, being closely linked with food production. Analysis shows that with increasing food production, consumption has increased; currently one-third of the global land area is used in food production. A trend of conversion of other land specially forest land for agriculture is already noted in South East Asia, specially in the hilly terrain.

Table 9.6 : Population in South East Asia

Demographic Variable	Country	Data	Source/Notes
1	2	3	4
Density (population/ sq. km.)	Asia - Southeast	128	Source: PRB 2007 World Population Data Sheet
	Afghanistan	49	Source: PRB 2007 World Population Data Sheet
	Bangladesh	1,035	Source: PRB 2007 World Population Data Sheet
	India	344	Source: PRB 2007 World Population Data Sheet
	Malaysia	82	Source: PRB 2007 World Population Data Sheet
	Maldives	1,020	Source: PRB 2007 World Population Data Sheet
	Myanmar	74	Source: PRB 2007 World Population Data Sheet

1	2	3	4
	Pakistan	213	Source: PRB 2007 World Population Data Sheet
	Sri Lanka	306	Source: PRB 2007 World Population Data Sheet
	Thailand	128	Source: PRB 2007 World Population Data Sheet
Population Mid-2007	Asia - Southeast	574,000,000	Source: PRB 2007 World Population Data Sheet
	Afghanistan	31,900,000	Source: PRB 2007 World Population Data Sheet
	Bangladesh	149,000,000	Source: PRB 2007 World Population Data Sheet
	India	1,131,900,000	Source: PRB 2007 World Population Data Sheet
	Malaysia	27,200,000	Source: PRB 2007 World Population Data Sheet
	Maldives	300,000	Source: PRB 2007 World Population Data Sheet
	Myanmar	49,800,000	Source: PRB 2007 World Population Data Sheet
	Pakistan	169,300,000	Source: PRB 2007 World Population Data Sheet
	Sri Lanka	20,100,000	Source: PRB 2007 World Population Data Sheet
	Thailand	65,700,000	Source: PRB 2007 World Population Data Sheet
Population 2025 (projected)	Asia - Southeast	691,000,000	Source: PRB 2007 World Population Data Sheet
	Afghanistan	50,300,000	Source: PRB 2007 World Population Data Sheet

1	2	3	4
	Bangladesh	190,000,000	Source: PRB 2007 World Population Data Sheet
	India	1,391,200,000	Source: PRB 2007 World Population Data Sheet
	Malaysia	34,500,000	Source: PRB 2007 World Population Data Sheet
	Maldives	400,000	Source: PRB 2007 World Population Data Sheet
	Myanmar	55,400,000	Source: PRB 2007 World Population Data Sheet
	Pakistan	228,800,000	Source: PRB 2007 World Population Data Sheet
	Sri Lanka	21,200,000	Source: PRB 2007 World Population Data Sheet
	Thailand	70,200,000	Source: PRB 2007 World Population Data Sheet
Population 2050 (projected)	Asia - Southeast	778,000,000	Source: PRB 2007 World Population Data Sheet
	Afghanistan	81,900,000	Source: PRB 2007 World Population Data Sheet
	Bangladesh	231,000,000	Source: PRB 2007 World Population Data Sheet
	India	1,747,300,000	Source: PRB 2007 World Population Data Sheet
	Malaysia	40,500,000	Source: PRB 2007 World Population Data Sheet
	Maldives	500,000	Source: PRB 2007 World Population Data Sheet
	Myanmar	58,700,000	Source: PRB 2007 World Population Data Sheet

1	2	3	4
	Pakistan	295,000,000	Source: PRB 2007 World Population Data Sheet
	Sri Lanka	19,500,000	Source: PRB 2007 World Population Data Sheet
	Thailand	68,900,000	Source: PRB 2007 World Population Data Sheet

Scientists believe that habitat loss, modification and fragmentation are most important causes for loss of biological diversity. Rate of extinction is likely to change and increase as a function of the area of habitat that has already been destroyed and it can be as high as 8 times more species with additional 1% of loss of habitat;(UNEP, 1995).

India alone has officially diverted 1,21,823 ha; of forest (Ghosh, 2007) between 1980-2003, after enactment of Forest Conservation Act in 1980; impact of such diversion on forest based biodiversity can be anyone's guess.

4.0 CLIMATE CHANGE

Sala and Chapin (2000) identified 'climate change' phenomenon as the second most important driver only next to land use change. It is expected that major effect of climate change on biodiversity will be at the high latitude. Impact of increased CO_2 on biodiversity is considered similar to that of nitrogen concentration. Studies indicate that significant differences may be noted among 'biomass', in case of future changes in biodiversity. Impact of climate change will be highest in Arctic, Boreal and Grassland system. In any case, global warming has now been established as a real happening when studies revealed that until 19th century the temperature was declining but as present it is warming up at a rate that is unprecedented in the history 'leading to the temperatures in the date 20th century that were unique to the context of the entire millennium '(Bradley, 2000).

In a recent paper, (Chapin III et. al., 2000 *Nature*, 405; 234 – 242) it is stated that human alteration of global environment has triggered the sixth major extinction event in the history of life and caused wide spread changes in the global distribution of organisms.

Major changes can occur in global vegetation cover in response to global climate change, primarily as a result of changes in temperature and precipitation (Gates 1993,). Scientists predict decline in boreal forests but increase in grassland and desert area. But complexities arise when one considers multiple interactions between temperature, precipitation, soil and increase in atmospheric carbon dioxide. Behavior of individual species and biological communities is apt to change in unpredicable way. The rate of both species invasions and extinction? Is likely to accelerate.

While ability of animal species to move out is more accepted, the ability of plant species to shift their ranges in response to climate change depend on dispersal mechanism of seeds and adaptiveness. It may be remembered that 80% of flowering plants depend for pollination and seed dispersal on insects and birds; migration of such species may lead to serious crisis in the life of plant community.

Most vulnerable will be (Peter & Lovejoy, 1992,):

- Population living near the edge of species range
- Endemic and Endangered species that exist only in reserves or extremely limited habitats.
- Species facing over exploitation and habitat loss.
- Species in alpine and montane environment due to altitudinal shifts triggered by increased temperature.
- Migratory species following seasonal change pattern and food availability.
- Animals prone to thermal stress and temperature dependent sex ratio (Fish, Reptile, Insects).

South East Asia has many endemic and threatened wild life species currently in reserves. With geographic range change of many plants, biomes may have serious impact on wild life. Due to reduction in current range change future of such species may be at stake.

While negative impacts are foreseen in most of such species, vector species like mosquito may breed faster resulting in spread of malaria and dengue in the region.

5.0 DISCUSSION

Global biodiversity is facing a crisis of survival in the present time. It is estimated that 20% species of mammals and 10% species of birds are currently threatened by extinction which is 100-1000 times greater than background level, (Chapin et. al, 2000). The Asian region perhaps faces a greater challenge conserving biodiversity. During the history of civilization over last 3000 years, the changes in land use over South and East Asia have been one of the largest in the world. (Kabat et.al. 2000, *Global change Newsletter*, 44:9-11). Wildlife Conservation Society during COP 7 pointed out to yet another critical issue of Wildlife and Biodiversity Conservation; WLS report states; in last 40 years, 12 species of large animals have become extinct in Vietnam; in India tiger population is now facing the greatest crisis. Wildlife trade in China has been widely publicized. As animals that play vital role in the forest as predators, pollinators and seed dispersers disappear due to uncontrolled land use changes, fragmentation of the habitat, change of vegetation and forest cover and illegal trade, other species are likely to become rare or extinct. Original habitat for biodiversity in 19 countries of Asia has been reduced by 68% (McKinnon & McKinnon, 1980, IUCN).

It may be noted that the studies didn't include marine ecosystem based biodiversity which is also undergoing serious changes due to unprecedented events like "Tsunami" and is likely to face critical change due to global warming specially in planktonic diversity. Studies have shown that large differences in CO_2 sensitivity between phytoplankton species. A shift in phytoplankton species composition and succession is likely to impact both ecosystem regulation and biochemical cycling (Riebesell et. al, 2000, *Nature*, 407: 364-367).

It is highly speculative to predict the exact extent of change that may occur due to land use changes, climate change and increasing human population. Undoubtedly, the changes in biodiversity in the 21st century would be serious, critical and deserve attention beyond verbal rhetoric. While the "Convention on Biological Diversity" 1992, was adopted by the Global community, at the right time, unless appropriate actions are taken the objectives of conservation would never be fulfilled. It's not only rate of extinction of wild species but also the declining population of many endemics that are causing concern. The eroding agro biodiversity with the extension of industrial agriculture and vanishing indigenous breeds of livestock due to so called modern animal husbandry demand serious attention and corrective actions. Regional co-operation in conservation of biodiversity has remained largely in paper; collective action can lead to collective gain. Countries of ultimate responsibility in South East Asia must act together.

It is said that 8000 years ago entire South East Asia was forested. In 21st century the forest cover has dwindled alarmingly specially between 1800 – 2000 period; some countries like Singapore has entirely denuded forest land. Conversion due to agriculture is said to be more than 1 million hector annually; deforestation has been calculated to release 465 million tonnes of carbon to the atmosphere. Wildlife trade has risen sharply, 6 times beyond the level of sustainability rate. Hundreds of slow breeding animals, non-human primates, large carnivores, elephants are becoming vulnerable. Besides direct human interference forest fire tunes its own toll. It is indicated that 42% of biodiversity of South East Asia will be lost at the percent rate of deforestation by 2100(Archard et. al., 2002 in Sodhi and Brook, 2005).

The European Science Foundation (ESF) has set up an European Collaboration Research Programme, which include probing from past climates to future biodiversity (*Global Change Newsletter* no 66, October 2006). To understand 'fundamental ecological and social processes that drive biodiversity changes'-this project is expected to initiate serious research to assess human foot print worldwide. Such a project is yet to be conceived in South Asia or in Asian context, to validate experimental and empirical data available from the region.

REFERENCES

Bradley, Raymond 2000. 1000 years of climates change. *Global Change Newsletter*, 44: 5-6.

Gates, D. M. 1993. Climate change and its Biological Consequence. Sinauer Associates, Sunderland, Mass.

Ghosh, A K 1995. Biodiversity in South Asia 112 – 125. In "Emerging South Asian Order, Hopes and Concerns." (eds. Sankar Ghosh and Somen Mukherjee). Media South Asia, Calcutta, India.

Ghosh, A K 2005. Asian Biodiversity Conservation. *Asian studies*, 23 (182) 27 – 36.

Ghosh, A K 2007. Environment; Issues and Concern. APH, New Delhi.

Johnson, Nels, 1995. Biodiversity in the Balance: Approaches to setting Geographic Conservation Priorities. Biodiversity Support Programme World Wild life Fund, Washington D.C., USA.

Jolly, C. L. and Tarry B B (eds) 1993. Population and Land use in developing countries. National Academy Press, Washington D.C., USA.

Mackinnon, J. and Mackinnon, K. 1986. Renew of the Protected Areas System in India Malayan Realm. IUCN, Gland, Switzerland.

Myers, N. 1988. Threatened biotas: 'Hotspot' in tropical forests. *Environmentalist* 8(3): 1 – 20.

Myers, N. 1990. The Biodiversity Challenge: Expanded 'Hotspot' analysis. *Environmentalist* 10 (4) 243 – 256.

Peters, R. L. and Lovejoy, T. E. 1992. Global warming and Biological Diversity. Yale University Press, New Haven, Conn., USA.

Riebesell, U. Zonderven, B. Rost and Zeebe, R. E. 2001. Effects of increasing Atmospheres Co2 on phytoplankton communities and biological carbon pump. *Global Change Newsletter*, 47: 12-15.

Sala, Osvaldo, E. and Chapin, Terry. 2000. Scenarios of Global Biodiversity. *Global Change Newsletter*, 43: 7 – 11, 19.

Sala, Osvaldo, E. 2000. Global Biodiversity scenarios for the year 2100. *Science*, 287: 1770 1774.

Sodhi, N.S. and Brooks, B. W. 2005. South-east Asian biodiversity in crisis. Cambridge University Press, Cambridge, U.K.

Stuart, Chapin III et. al. 2000. Consequences of changing Biodiversity. *Nature*, 405: 234 – 242.

Tilman, D., May, R. M., Lehman, C. L., and Nowak, M.A.1995. Habitat destruction and extinction debt. *Nature*.

UNEP 1995. Global biodiversity Assessment. (ed V. H. Heywood). Cambridge University Press, Cambridge.

CHAPTER X

Sacred Groves

1.0 INTRODUCTION

Biodiversity Conservation has now been accepted as a government sponsored programme at the national and international level. The institutional support system for *in situ* conservation is restricted to the notified protected area (PA) network of wildlife Sanctuary, National Parks, Biosphere Reserves, Tiger Reserves, Elephant Reserves etc. More than 600 PA's now exist in India.

But beyond government control, an age-old ethics of conservation exists in India, developed and protected by faceless, nameless common people. Religion being one of the strongest cultural link in the country, the part of nature protected by people has often been linked with a 'sacred' cover. Sacred groves known by different names have been gradually documented, by researchers and with expanding knowledge base the richness of data emerged.

Sacred Groves prohibit hunting, removal of plants sometimes even wood and leaves. The vegetation cover help to maintain an ecological balance, protect microclimate, help to maintain soil nutrition and offer an unique habitat to the wildlife (Malhotra, 1998).

The word sacred Grove can be traced back to 1897 when Brandis (1897), the acknowledged founder of modern forestry in India, recognized this indigenous form of conservation. Kosambi (1962) argues its pre-agrarian existence, but strongly enough no serious research on the subject could be traced in mid 1970's.

Brandis (op.cit) wrote "Very little has been published regarding sacred groves in India, but they are or rather very numerous ..... These.... as a rule, are not touched by the axe, except when wood is wanted for repair of religious buildings."

2.0 MANY NAMES: ONE AIM

The sacred groves are called differently in different geographical locales, viz., *Than* in West Bengal and Assam, *Ka law kyntang*, *Ka law adong*, *Ka law lyngdow* in Meghalaya, *Sarna* in Jharkhand and Madhya Pradesh, *Devrahi* in Maharastra, *Oran*, *Vani kenkri* in Rajasthan; in the southern part *Kovil kadu* in Tamil Nadu, *Devarkdu* in Karnataka and *Kavu* in Kerala carry the same meanings. In a multi-lingual society, this is expected but the strong ethics of conservation emerges through such acceptance of conservation area designated by people.

A recent report shows that so far 58 sacred groves have been recorded in Arunachal, 40 in Assam, 365 in Manipur, 79 in Meghalaya and 670 in West Bengal, 322 in Orissa and 750 in Andhra Pradesh. In Northeast and eastern India alone as many as 2284 sacred groves have been identified under different names. In the tribal belts Jharkhand (27) and Chattisgarh (600) sacred groves have been recorded whereas in southern states of Tamil Nadu (448), Kerala (2000), Karnataka (1424) a staggering 3872 sacred groves are noted. In western Indian states of Maharastra (1600), Gujarat (29), Rajasthan (241) and Northern India, Haryana (248) are recorded. In Central India, Madhya Pradesh accounts for 21 sacred groves but up in the Himalaya, Himachal Pradesh presents the single largest number of 5000 sacred groves. Interestingly, no data from Uttar Pradesh, Uttaranchal and Jammu & Kashmir is available.

Added up, the figures stand at 13,922 sacred groves as against 4875 estimated by Malhotra et.al. (1999). Needless to say, the figures only represent the extent of investigations; for instance, the number in Madhya Pradesh and Jharkhand appears too low. Although the figure from Sikkim does not feature, the highlands of Demojong below Khangchendzongh peak is said to be the most sacred site (*Down to Earth* Supplement No. 3, December 31, 2003).

3.0 BIODIVERSITY IN SACRED GROVES

Ramakrishna et.al. (1998) edited an exhaustive volume aptly titled 'Conserving the sacred for Biodiversity Management'. It was Gadgil and Vartak (1976) who should be credited to present perhaps the first consolidated account of Sacred Groves of identified geographical area globally well known for its rich biodiversity, the Western Ghats. Socio-historical studies and socio-cultural researches have been interlinked with the heritage of sacred groves. Sacred groves, by virtue of protection by local community not only conserve relict flora but act as habitat for evolution of new species; new species of insects disclosed from the sacred groves near Mawphlong in Khasi Hills of Maghalaya or frogs from Amarkantak, Central India bear evidence to such findings. Conserving rare and endangered species not by Indian Wildlife Act or other legal provision but restrictions put by the local community remains a focal point. The role of sacred Groves as refugia of rare and threatened species was first highlighted by Joshi and Gadgil (1991). Subsequently Deb, Deuti and Malhotra (1997 a) showed the role of sacred grove as bird refugia.

The endangered plants located in different Sacred Groves in India include *Flacurtia cataphrails*, *Kunslerea keralaensis*, *Phqeba hainsiana*, *Rhus hookeri*, *Syzygium travancorium* to name a few. In Andhra Pradesh, endemic *Pterocarpus santalinus*, *Shorea tumbaggaia*, *Terminalia pallida*, *Cycas beddomei* and many others in the Sacred Groves of Chittor district. Ancient trees of 18-30 m height and 7.5-11 m girth of 200-300 years old were seen in Cuddapan district.

However, value of Sacred Groves was never appreciated by the western colonial rulers. Ignoring the cultural and heritage value of these unique patches of forest, vast areas specially in Eastern India underwent drastic changes. Permanent Settlement Act, 1793 and later Forest Act, 1876 led to drastic change of land use for a centralized government control system. Even then sacred groves continued in all provinces. Deb (unpub.) rightly pointed out the historical pathway of declining sacred groves from the days of Brandis (1897) to the Independence of India (1947). Strangely enough, the Sacred groves in the Southern states survived through the period of political changes.

Gadgil and Vartak (1981) wrote "these sacred groves are now playing the role of reservoirs of biological diversity. It is well known that climax forest land lends to be richer in number of plant species than other stages of succession. As few of the remaining examples of climax vegetation the sacred groves often contain more plant species .... it is important to preserve this diversity for posterity".

Researches on Sacred Groves actually thrived only after 1980's. One can locate more than 40 interesting research publications during next two decades; synopsis of which can be reviewed in the volume edited by Ramakrishna et.al. (1998) and Malhotra et.al. (1999).

4.0 MANAGING SACRED GROVES: FROM MEGHALAYA TO KARNATAKA

In the northeastern state of Meghalaya, Sacred groves have often been not isolated from the main activity centre of the community. Concept of demarcating core and buffer zone is unknown but an unwritten code of ethics prevails. In some cases, even shoes are to be taken out and walking barefoot in the abode of god is the norm. In an atmosphere of reverence, an array of rich flora – with nearly 50 per cent being rare can be located. Entering such forest may require permission of the local tribal head. In Meghalaya, the groves are differently named like, forests under traditional religious head, forests protected for non material use, forests for village use and forests most sacred etc. Such groves may be as large as 100 ha., or as small as 0.1 ha. Since forests in Meghalaya is owned largely by the local community, the management of the groves rest with the local people.

The situation in eastern India, specially in West Bengal may present a different profile. Small sacred groves are often protected by the local community, but one of the last relict forest in the Vast Ranigunj – Asansole coal belt is governed by the forest department. Obviously community protection in the plains largely depend on funding support. In Orissa, many groves in tribal belts are managed by clans of tribes.

Management of Sacred groves in Maharastra have largely been taken over by the State Forest or State Land Revenue Departments. In most of the 2808 sacred groves in the state cover less than 1 ha., and are locally called *Devrahis* or *Devrathis*.

It is reported that community-owned groves in Maharastra have suffered the most because of peculiar policy of land revenue department: community forest is to be registered in the name of any selected individual (say village priest). In tribal belts, some groves are still 'owned by the earliest settling clans or families' (Viswanath, *Down to Earth*, 2003 a). Sacred groves often changed its ritualistic characters of ancient tribals to the temple-oriented religion. The Pashim Maharastra Devasthan Samiti is more eager to protect temples than the surrounding forests and biodiversity. Such changes in priorities are distinct evil signals.

Rajasthan, in spite of its desert environs and scarce forest cover has become well known for conservation ethics followed by the Bishnoi Community. The *Oran*, *Vani*, or *Kenkri* in local language are fiercely protected by the community. Poaching in these are totally banned. Sacred areas are used for grazing, collection of biomass, medicinals but never used for collecting timbers. Nine sacred groves across 241 ha., of land has so far been documented.

Biswanath (*Down to Earth*, 2003 b) has given an alarming account of loss of sacred groves in Kodagu area of Karnataka. It is surprising to note that 873 groves covering 4,398 ha., were listed from Kodagu as far back as in 1873 (perhaps the earliest systematic study of sacred groves in India). With enactment of Indian Forest Act in 1875, these were declared as "protected forests". The onslaught of government system didn't impact the characters of these groves and as a matter of fact, a report of Commissioner of Coorg (Kodagu) dated 1898 recorded 6,275 ha., as sacred grove out of total of 45,0118 ha., forest area. But in 1905, the management of these *Devarakadus* went to the Revenue Department. And between 1905 – 1985, the sacred grove area declined by 42 per cent to 2406 ha.! In between, the groves were transferred to the State Forest Department under Indian Forest Act, 1927 but the squabbling between Revenue and Forest Departments took its toll. In 1985, the Revenue Department formally handed over the ownership of *Devarakadus* to the Forest Department, and these were declared as "Reserve Forests". by 1990's, 45 per cent of the groves had been reduced to less than one acre; between 1898-1990, the sacred groves of Kodagu declined from 14 per cent of forest area of the state for merely 2 per cent.

Even then researchers believe, Kodagu's sacred grove *Devarakadus* got a better deal than similar groves *Kans* called in Simoga district. Forest Department never recognized, *Kans* intrinsic value. These have been declared as minor forest, state forests or reserve. The *Kans* occupy 5.85 per cent of the total land area of Uttar Kannada. And disregarding the cultural and conservation value, State Forest department even planted *Acacia* in the *Kans* during 1966-1985!

5.0 CULTURAL AND SOCIAL INTEGRATION

The sacred groves are believed to have evolved through a tribal culture of worshipping nature. The researches carried out in tribal belts of Northeast India and also in the plains and hills of West Bengal, Jharkhand, Orissa, Chattisgarh and Andhra bear testimony to the same (Deb and Malhotra, 1997 b). The changes from typical tribal ethos to the Hindu or Muslim religion based tradition of protecting nature as sacred grove however has not been uncommon. Researches by Deb (NBSAP Doc., unpub.) have shown groves of tribal origin could be discerned by tribal nomenclature even when tribal and subaltern traditions are brought under Hindu pantheon. The classic example of a Brahmin priest worshipping *garam-than* (sacred roves in Santhali language) without a deity is one such classic example from Puruliya district of West Bengal. Deb (op.cit) also recorded a number of Muslim documented sacred groves.

Analysis of the culture of sacred groves therefore shows that conservation, in the name of religious, belief cutting across the tribal non-tribal divide or Hindu-Muslim divide remained the key word. Common people didn't know the value of rich generic researches but was aware of their use values at the hour of need. Sacred groves not only protected the flora with religious or totemic values but plants of no known religious value found safe shelter.

6.0 THE CRISIS OF EXISTENCE

In spite of extensive publications on Sacred Groves during 1980-2000 and beyond, these unique features however fragmented across India, are fighting a battle for survival. Transformation of sacred groves in Karnataka is just but one example; conversion of sacred groves can never be compensated by compensatory afforestation. Loss of the oldest banyan tree (*Ficus bengalensis*) near Kadiri in Anantapur, Andhra Pradesh will be irreplaceable. The tree has grown and the villagers graciously given their land to help it grow as the largest banyan in the World (Anon, 1996, WWF-I). Timber from sacred groves are often used for building temple; the ground cover in the forest of climax vegetation, when removed not only leads to loss of soil nutrition but also in disturbance of surface flow and recharging of groundwater.

The problem of sacred groves also are emanating from the so called modernization of thoughts in tribal culture. Young members of the society in many parts of the tribal India consider the practice of sacred groves as relicts of superstition. This dangerous trend of 'erosion of belief system' was noted by Dashi (1992) and Viswanath (2003).

The lack of protection, lack of authentic geo-referenced data base, change in tribal belief and above all varied system of land ownership and development, collectively endanger the tradition of sacred groves in India today.

REFERENCES

Anon, 1996. Sacred and Protected Groves of Andhra Pradesh. WWF-India, Hyderabadand New Delhi.

Brandis, D., 1897. indigenous Indian Forestry. Oriental Institute, Woking.

Dash, S., 1992. A tradition of transformation. p. 112. *In* Tribal India: Ancestors, Gods and Spirits (ed. Sanju Dashi). Marg Publication, National Centre for Performing Arts, Bombay.

Deb, Debal and Malhotra, K. C., 1997 b. Interface between biodiversity and tribal cultural heritage. *J. Hum. Ecol*, 8: 157-163.

Deb, Debal, 2002. Sacred Groves. Document for NBSAP (unpublished).

Deb, Debal, Deuti, K., and Malhotra, K. C., 1997 a. sacred groves relics as bird refugia. *Curr. Sci.*, 73: 815-817.

Gadgil, Madhav and Vartak, D., 1976. Sacred Groves of Western Ghats of India. *Economic Botany*, 30: 152-160

Gadgil, Madhav and Vartak, D., 1981. Sacred Groves of Maharastra: an inventory. *In* "Glimpses of Ethnobotany" (ed. S. K. Jain). Oxford University Press, Bombay.

Joshi, N. V., and Gadgil, Madhav, 1991. On the role of refugia in Promoting Prudent: Use of Biological Resources. *Theoritical Population Biology*, 40: 211-229.

Kosambi, D. D., 1962. Myth and Reality. Studies in the formation of Indian Culture. Popular Prakashan, Bombay.

Malhotra, K. C., 1998. Anthropological dimensions of sacred groves in India. p. 423-438. *In* "Conserving the Sacred for Biodiversity Management" (eds. P. S. Ramakrishna, K. G. Saxena and U. M. Chandrasekharan).

Malhotra, K. C., Gokjale, Y., Chatterjee, S., and Srivastan, S., 1999. sacred Groves in India: An Overview. National Museum of Mankind, Bhopal.

Ramakrishna, P.S., K.G. Saxena, U. M. Chandrasekharan, 1998. Conserving the Sacred for Biodiversity Management. Oxford and IBH Publication, New Delhi.

Viswanath, Ranya, 2003 a. From Groves to temples. p. 71-75. *Down to Earth*, Supplement No. 3, December 31, 2003; (ed.) An Inflected Landscape', p. 76-78. *ibid.*

CHAPTER XI

Wildlife Crisis

1.0 GENESIS OF WILDLIFE CONSERVATION

The conservation issue related to wildlife in India dates back to the colonial period. Post Independence era witnessed the first concerted efforts of the Government of India in the formation of Indian Board of Wildlife as far back as in 1950. With myriads of problems facing the divided country, it was heartening to see the institutional support system to consider the future wildlife in India.

Since then remarkable progress has been achieved to strengthen the efforts. The enactment of Wildlife Protection Act and Rules in 1972, the year the first World Conference on Environment and Development was held in Stockholm, Sweden, ushered a new era. Indian wildlife was protected on the basis of survey data in different categories viz., Endangered, Threatened, Rare and Vulnerable, enlisted in the schedule I to V (later a schedule VI was added for only 6 species of endangered plants). Strangely enough, "wildlife" was classically confined to the faunal species, and largely the same concept persists. The act laid down the procedure for appointing authorities (Director, Chief Wildlife Warden and other officers) and constitution of Wildlife Advisory Board at the State level; it also laid down procedure of banning hunting, protection of wild plants (Chapter III A), declaration of Sanctuaries, National Parks, specified constitution of Central Zoo Authorities (Chapter IV A), restriction on the trade or commerce in wild animals, animal article, trophies (Chapter V), prevention and detection of offences (Chapter VI) etc. (Anon, 1994). The act has since been amended from time to time, the last revised and amended version has been notified on 2nd April 2003.

Forest and Wildlife have been included in the Concurrent List of Constitution by the 42nd Amendment in 1976.

India has also become a party to the International Convention on Trade in Endangered Species of Wild Flora and Fauna in 1976. To control the trade in endangered species listed in the Appendices of the Conventions offices have been set up in four metro cities. India is also a member of Ramsar Convention, Convention on Migratory Species (CMS).

2.0 PRESENT STATUS

In 1980, India had 19 National Parks and 205 Wildlife Sanctuaries in 26 States and Union Territories (Anon, 1981). In 2006, the number has increased to 92 National Parks and 500 Wildlife Sanctuaries covering 15.67 million hectares (MoEF Annual Report, 2004 – 2005). In addition, 14 Biosphere Reserves have been declared covering an area of 55,771.95 sq.km., in almost all major ecosystems. Additionally, 38 mangrove forest areas in 10 states and four coral reef areas viz., Andaman and Nicobar Islands, Lakshadweep Islands, Gulf of Kutchch and Gulf of Mannar are provided intensive conservation since 1987. A total of 28 Tiger Reserves in 17 states covering 37,761 sq.km., have been set up at different time period. Project Elephant launched in 1992 is being implementes in 12 states.

The Government of India has covered 31 rivers in the country under River Conservation Programme; National Lake Conservation Programme is supporting restoration and conservation of 28 lakes while 71 wetlands in 22 states have now come under a National Wetland Conservation Programme. As many as 25 wetlands have been declared as Ramsar Sites in India. the list undoubtedly is very impressive; the efforts in conservation of wildlife and biodiversity (*in situ*) during 1972-2006 or little over three decades demand commendations (Annual Reports, MoEF, 2003 04, 2004-05, 2005-06).

3.0 AREAS OF CONCERN

3.1 Tiger count

While efforts have been commendable, the management strategy for wildlife conservation appears confusing in more than one areas. Take for instance, the international support India had received when it launched Project Tiger as a Central Sector Scheme in 1973 in nine tiger reserves. The network of Tiger Reserves has since then increased significantly but a deep crisis in tiger conservation has been revealed since 2005. Questions have been asked on the actual count of Tiger, the method adopted for census operation and aspersions have been made on the pressure on forest officials managing Project Tiger to show ever increasing number. The real life story of Sariska and Ranthambore created enough stock and awe. Even Prime Minister had to intervene. Media reports rightfully asked the rationality of expenditure on "Project Tiger" when the revelations become too shocking.

This was followed by the apparent controversy in number of Tigers in the prime habitat of the Sundarbans. Indian Statistical Institute revealed that the census data published by the Forest Department is exaggerated, the real number fall far below. Public allegations and counter allegations covered media reports in 2006.

Till date, people of India have no authentic data on this keystone species. Crores of tax payers money and international fund have been spent. Obviously poaching and trade data remain fuzzy. Blame on Chinese trade does not absolve the authorities set up for protection Indian Tigers. In 2005-06, only a sum of Rs. 2875–97768 was released for 17 Tiger Range States of India; the allocation under 9th Five Year Plan was Rs. 7500 lakhs, it was doubled during 10th Five Year Plan to Rs. 15,000 lakhs but the Tiger population continue to decline!

The Project Elephant was launched to protect yet another endangered large mammal. On an average Rs. 12 crores / 13 crores is being allocated every year for conservation of elephants. But human – wildlife Conflict seems to be a major issue in elephant conservation which remains largely unresolved specially the western plains of West Bengal. Habitat destruction in Dalma Forest neighbouring Bihar (now Jharkhand) is said to be the root cause of annual migration to West Bengal, largely in search of food. Since the forests in Puruliya, Bankura and West Medinipur districts of West Bengal can not provide the requisite food plants for elephants, frequent reports of elephants straying into village area are published in the media; Forest Department of West Bengal reported death of 21 adult male elephants due to poaching and 9 adult elephants due to accidents between 1996-97 to 2002-03. 74 people have been killed by wild elephants – 20 in South Bengal and 54 in North Bengal in one year, 2002-03 (Central Data Bank, Wildlife Wing, Forest Directorate, Government of West Bengal, 2002-03). Similar data from other states however, are not easily available. But that the "Project Elephant" has failed to resolve Human – Wildlife crisis is clear. It has also failed to protect this schedule I animal in many parts of India e.g., Karnakata. It is reported that India's elephant population has gone beyond the carrying capacity of its habitat area.

The impact of globalization on Gross Domestic Product (GDP) not withstanding, the accelerated mining activity in prime habitat area can only be expected to increase Human – Wildlife conflict and make the species more vulnerable in Jharkhand and Chattisgarh. The conflict between Environment and Development is increasingly becoming visible in such case studies across India.

The Golden Triangle Project for building up super-fast roadways has already been in progress. While appreciating needs for world class infrastructure development, the planning of expansion of existing roadways through prime wildlife areas often create serious concern. With much debate, the realignment of NH-34 in North Bengal (which passes through all major conservation areas) in sub Himalayan eastern India, has been made, such process of planning can lead to devastating effects in wildlife habitat.

Likewise the project on "Interlinking of Indian Rivers" in the Himalayan sector did not pay attention on the serious impacts of prime wildlife area in Assam, North Bengal and several other hill-forest areas. National protest have not been able to make the Government concede to the demands for full environmental impact assessment by competent expert till date.

Developmental demands have already led to de-notification of prime sanctuary areas in North West Himalayan state of Himachal. The de-notification of parts of Majathal Wildlife Sanctuary for Kol H.E. Project, Darlaghat Wildlife Sanctuary for Ambuja Cement and Narju Sanctuary in Mandi for road construction bear testimony to the violation of the principle of *in-situ* conservation of wildlife. Earlier attempts to denotify Narendrapur Wildlife Sanctuary near Capital city of Kolkata was only foiled after 15 years of sustained campaigns by city based NGO's.

3.2 Marine Protected Area

India has declared two marine biosphere reserves viz., Gulf of Mannar and Sundarbans and listed more than 20 marine protected areas. These strangely enough are managed by the State Forest

Department; that forest ecosystem and coastal and marine ecosystem demand separate line of expertise is undeniable. As a result, the conservation programme both in terms of survey and documentation, monitoring of habitat degradation, population status suffer significantly.

India deserves to have more marine protected areas (PA) in 7500 km of coastline and 2 million square km., of exclusive economic zone. The interaction between acknowledged marine institutions e.g., Central Marine Fisheries Research Institute, National Institute of Oceanography, Centre for advance Studies of Annamalai university, coastal universities, Marine Biological Stations of Zoological Survey of India and the authorities of marine PA's remain almost insignificant. India needs a separate "Marine Protected Area Authority" to be manned by marine scientists under the Department of Ocean Development or Ministry of Environment and Forests for proper management system. In the meanwhile, crisis continues.

3.3 Data and Management Plan

Wildlife conservation immensely suffer from lack of *authentic* data and in many instances basic benchmark data needed for assessment of lacking; for instance, India Eco development Project (IEDP) aided by The World Bank was meant to augment biodiversity and conservation by reducing human impact through livelihood improvement programme; baseline data needed to assess the impact being absent, no proper evaluation was able to be undertaken in many of the project areas.

Management Plans for PA are vital for implementing wildlife action plan. However, either management plans are delayed or even when prepared suffer insights into the problem area. Post implementation evaluation of management plan remains hasty and may not reveal the impact of investment vis-à-vis return. This is more evidenced in the areas of Biosphere Reserves and Buffer Zones with human habitation.

Multilateral support to a single programme in the wildlife habitat improvement lacks coordination and accountability. These are common syndromes in many parts of forest and wildlife management.

Researches on wildlife remain mostly confined to the Universities, NGO's or Institutions like Wildlife Institute of India

(WII), SACON etc. Research Wings within wildlife division often suffer from lack of wildlife biologists and taxonomists – both vital for identification, documentation and biological researches on food availability, carrying capacity and other basic areas.

While WII and other well known academic centers and NGO's like BNHS have produced excellent, trained, young wildlife biologists, their deployment in wildlife division remain very restricted due to dearth of adequate career promotion, lack of infrastructural support like laboratory, equipment and vehicles.

3.4 Wildlife Trade

India became well known due to poaching of wild elephants by Verrapan and group in the South and such traders who continued to escape law – like Sekhar Chand in Delhi. Indian Wildlife Trade data is collected by officials of the division, Wildlife Directorate of Government of India and by the volunteers of TRAFFIC. Coordination between agencies continues to be fragile and incoherent. Products made of banned wildlife parts became difficult to identify in many cases; inter-state and inter-country trade in wildlife often feature in the media. Perhaps the worst victim of wildlife trade in recent years is the Royal Bengal Tiger in Indian Forests. Trade in Ivory, crocodile and tiger skin, skins of clouded leopard and leopards and other big cats and trade in Rhino horns have features periodically but the actual extent has never become clear.

3.5 Wildlife Tourism

India has become a much favoured destination for wildlife tourism. However, Infrastructure support to promote ecotourism is still a far cry. Tourists are hardly apprised on the value of the wildlife in human life or oriented towards good practices during their visits. Most of the PA's can't provide basic information, picture postcards, booklets, souvenirs, CDs on bird calls, maps and trek routes. These are essential items for any ecotourism programme. Sadly, with vast potential, interpretation centers even when set up, remain in a wrong site, not at the entry point.

Trained wildlife guide is a prerequisite for wildlife tourism. Involvement of local people, after imparting necessary training, has been proved to be one of the best methods for such a programme.

Training programmes for guides in wildlife tourism, based on specific location of PA's, is an essential part of successful strategy. Roads, Accommodation, Water supply apart from the need of attractive material, souvenirs, pictures, CDs can make the venture a real success, aiding conservation.

4.0 DISCUSSION

The issues raised in the previous sections reflect some of the crisis areas for wildlife conservation programme in India. Admittedly, the efforts to expand the network is laudable but many of the essential support system is yet to be in place. Wildlife conservation could one day be a self-financing project in forestry sector but it demands a holistic approach for management, accountability and sincerity. The expected participation of the members of Forest Protection Committee (FPC) or Eco Development Committee (EDC) is protecting wildlife remain a myth since the villagers do not have means to defend themselves or access to facilities to transmit the message of wildlife crime.

NGO's can play a more significant role but very often the real coordination in the form of participatory management remains a far cry.

India needs a re-look at the current wildlife action plan. Otherwise, one may be rightfully apprehensive of a time when the wonder that was Indian Wildlife may become a topic for conservationists.

REFERENCE

Anon, 1981. Wildlife in India. Department of Agriculture and Cooperation. Ministry of Agriculture, Government of India.

Anon, 1994. The Wildlife Protection Act, 1972 as amended up to 1991. Nataray Publications, Traffic India.

MoEF, 2003-04. Annual Report. Ministry of Environment and Forests, Government of India.

MoEF, 2004-05. Annual Report. Ministry of Environment and Forests, Government of India.

MoEF, 2005-06. Annual Report. Ministry of Environment and Forests, Government of India.

CHAPTER XII

Invasive Species

1.0 INTRODUCTION

The global biodiversity conservation has now been accepted as a priority agenda. The Convention on Biological Diversity (CBD), 1992 has an impressive list of countries as contracting parties pledging to save the rich living resources sustaining the ecosystem and human society. While countries have been sensitized over necessity of conservation of species, the ever-larking shadow of 'invasions' looms large (Dreke, J. A. et.al., 1989). CBD has stressed under Article 8(h), to 'prevent the introduction of, control, or eradicate those alien species which threatened ecosystem's habitats or species'.

Alien species make way to another territory either through planned introduction like Rubber or Coffee in India and elsewhere or by accident, international trade and travel. Accidental introduction of a species may lead to invasion of the species into new ecosystems and can cause havoc disrupting the ecosystem processes. Planned introduction based on serious research, on the other hand, helps in controlling pests and diseases or increase economic activities; Europeans brought the horses along with the house cats, dogs, cattle, sheep, goats years back and all of these have been of immense value to the society. Invasive species can cause loss of native species like introduction of Chestnut Blight from Europe led to rapid and dramatic loss of Chestnut in eastern deciduous forests of North America (GBA, 1995, p. 448). However, it had 'no discernible consequences for ecosystem functioning as the other tree species seem to have fulfilled its original functional role' (GBA, op.cit). But the scenario elsewhere may provide a drastically different impact profile, e.g., *Mikania* and *Eupatorium* invasion in Northeast India. introduction of red deer to New Zealand and rabbits to Australia actually caused serious problem and then the foxes were introduced to control them. One of the striking and destructive introduction by human has been the 'brown tree snake from areas of South Pacific to Hawaii, where it has devoured many species of birds that have no chance to escape the hungry advances' (Lapointe, 2006).

Often, a species introduced may remain in small population waiting for the opportune moment when specific food may become available. Surviving below threshold level of population over long period of time give credence to the species but when it gets the opportunity it may spread like a wild fire.

The invasive species may include an array of species from the group of Virus, Microorganisms to Plants and Animals. Other than planned introduction, the actual time and period of accidental entry is hard to determine. However, after adoption of CBD, extensive researches have been initiated. Electronic search for information on invasive species could lead the reader to 20 million information sources, but sadly information on India is largely lacking.

2.0 INVASIVE SPECIES IN INDIA

The only database on invasive species in India includes 108 scientific names; of these 59 belong to diverse faunal groups, five to fungi and 44 to floral groups. However, the database prepared by the Informatics Division of National Chemical Laboratory (NCL), Pune, could lead to country / region of origin for only 77 species. Obviously, the list is far from being complete as it lacks total information on Viruses, Monera as also information of many well known alien pests.

The list produced by NCL can be compared with the one provided by Pimental (2001) on total and alien species recorded from India (Table –12.1). However, not all alien species may be invasive and many are restricted to specific geographical area.

Table – 121: Alien Species in India

	Total	Alien
1	2	3
Plants	45,000 *	18,000
Mammals	316	30

1	2	3
Birds	1221	4
Reptiles & Amphibia	741	NA
Fishes (Freshwater)	2546 **	300
Arthropods	54,430	1,100
Microbes	NA	NA

* The number now stands at 47,000 including 17,000 flowering plants species (Author)

** This seems to be erroneous as the total fish species (freshwater, brackish and marine) from India approximates the same number (Author)

[Source: Pimental, David, 2001. Economic and Environmental Threat of Alien Plant, Animal and Microbe Invasion, *Agriculture, Ecosystem and Environment*, 84 (1)]

Species invading India have their origin in every major continents of the world extending from South America to Australia. Some familiar species of fishes e.g., *Cyprinus carpio* from Europe and *Gambusia* from America both of which have apparently proved economically beneficial to the country as has *Tilapia* from Africa. But *Cyprinus* is now known as destroyer of rooted aquatic vegetation which forms habitat for local species and act as feed for waterfowl. The removal of vegetation increases water turbidity, release locked up phosphorus and loss of micro invertebrates. *Gambusia affinis* is now known to predate upon eggs of economically important food fish and sometimes preys on indigenous ones. But others like the case of *Psidium guajava*, the guava introduced from Central America, Rubber from Malaysia or Coffee and Cocoa from South America which have all benefited agricultural economy.

But invasion of *Eichhhornia*, the water hyacinth from South America has created a massive problem of choking wetlands and waterways; the problem of North American *Potamegaton* in the waterways or even in the Chilika Lake of Orissa (when its salinity level dropped critically) or *Pistia* from USA in freshwater wetlands have cost money, labour and loss of productive income for fisher community.

The all pervading African Garden Snail is another alien invasive species, which has caused serious damages in plantations and home gardens in India. Till date, in spite of funded research project, eradication or control of population has not been possible.

The insect group Aphidoidea (Homoptera) includes largest number of species acting as virus vectors; otherwise, they can seriously damage economically important crops – from cereal to horticultural species. *Eriosoma lanigerum*, the wooly apple aphid was introduced from Europe to India and still remains a major pest for apples in the country. A cereal infesting aphid-vector and pest, *Hysteroneura setariae*, a typically North American species was first detected by this author in a old collection of aphid species from Lahore (now in Pakistan) dating back to 1914. The species remain undetected and unidentified till 1968 when colonies were identified on Rice Plants in Orissa; within two years, the species could be located along eastern coastal states, growing high-yielding varieties of paddy. This is classic case of an insect species which survived over 50 years below detectable level but became a true invasive species with changing agricultural practices. In recent time another insect, serpentine leaf minor (*Liriomyza trifoli*) introduced with Chrysanthemum cutting has spread to vegetable crops.

One of the two major biodiversity hot-spot areas of India has been besieged with one invasive species of plants *Eupatorium adenophorum* since 1970's. With heavy monsoon rainfall, the invasive species spread very fast in the whole of North Eastern region. *Eupatorium* has spread all over the roadside, hillocks and grazing grounds, in areas above 1000 m and even upto 2000 m (Barthakur and Ghosh, 1978). This is a pan tropical species, native to Mexico. A related species *Chromolaena odorata* (=*Eupatorium odorata*) has likewise invaded Indian plains, forming dense stands, which prevent other natural vegetation grow by competition and allelopathic growth (DTE, February 29, 2004). It is interesting to note that two alien species of insects were introduced in India to control these *Eupatorium* species viz., a gall fly from New Zealand for *E. adenophorum* and a lepidopteran from West Indies for *C. odorata*.

The other notoriously invasive alien species is *Mikania micrantha*, a South American plant reported to be imported in the early 20th Century in West Bengal, it has now spread all over Northeast India. It is considered the single most threat to forest, agriculture and tea plantation of the region. *Mikania* is most abundant between 100 m – 1000 m. popularly called mile-a-minute weed, this perennial weed prevent all other vegetation by cutting out the light (Choudhury, 1972; Datta, 1978). This unique quality of invasive growth is attributed to its phenotypic plasticity.

Such plasticity is also shown by other plant species like *Prosopis juliflora* in the semi-arid and arid zone of India, which was introduced from Mexico. Admittedly it has provided major economic return in terms of fuel wood use in the desert environ.

Another alien invasive species, *Lantana camara*, is now being researched by the scientists at University of Agricultural Sciences, Bangalore. *Lantana*, first introduced in 1907, is known to produce poisonous secretion through its roots, killing all other competitors. Taking advantage of its genetic plasticity, the plants produce flowers of different colours to take advantage of a wide range of insect pollinators.

In India, more intensive researches have been restricted to a selected plant species rather than on invasive animal species. Further, while microbial species introduction, particularly plant pathogen has been recognized to have significant effects on ecosystem composition in both natural and managed system, data on the same is largely lacking in India.

3.0 INVASIVE SPECIES: MANY FACETS

The invasive species introduced intentionally or accidentally, show several characteristics for successful survival and propagation in an alien land. Babu (2003) believes that only 10 per cent of invasive species population finds success in dispersal, of which 90 per cent fails to establish and of the remaining 10 per cent only 10 per cent could successfully spread. Invasive species is likely to face several barriers, geographical, environmental, reproductive and ecological. Absence of natural enemies helps the species to thrive in alien land. Attempts have been made to prepare predictive GIS based model for dispersal for a sugarcane aphid pest (Ganeshaiah, 2003), which indicate that due to ecological limitation, the pest may not be spread from Karnataka, Tamil Nadu, Kerala coastal areas.

The invasive plant species, according to Babu (op.cit.), must produce long lived seeds, must grow rapidly from vegetative to reproductive stage, should be capable of very high seed output in favourable environmental condition, seeds should be able to disperse over short and long distance and finally it should be strongly competitive. R. M. Uma Sankar, working on *Lantana camara* at the University of Agricultural Sciences, Bangalore (www.downtoearth.org.in) opined that invasive plant species would have high rate of success if it could attract insect pollinators and birds for dispersal. He has stressed that absence of associated herbivores in alien land could make the species more successful.

The invasive insect species in an alien land can successfully establish in absence of natural predators and parasitoids and in agreeable ecological condition. As the case of grain aphid has shown that such species can survive in low population density restricted in an ecological niche till favourable food plants become available.

Serious attempts have been made to search for biocontrol agents in their original homeland. But biocontrol agents need long time to establish and require successful mass culture for release at appropriate time. When in early 1960's, USA discovered spotted alfalfa aphid in western coast of California, search for natural enemies was extended to the original homeland in Rajasthan, India. The process is extremely time-consuming and cost intensive. Unlike in western countries, private investment in private insectaries is lacking in the developing world. Lack of opportunity to make easy profit (as in the case of production of chemical pesticides) acts as disincentive for investment besides constraints in limitation of technical knowledge and slow market uptake. Ramani and Bhumannayar (2003) in a mimeo publication (as quoted by DTE, February 29, 2004, p. 32) recorded 161 exotic natural enemies that have been studied in India as possible biocontrol agents; of these 69 were found viable but only 5 species gave significant results of the biocontrol agents; 55.5 per cent was tested to control aquatic weeds, 46.7 per cent was used to control homopterans (bugs) pests and 23.8 per cent was used against terrestrial weeds.

4.0 CONTROL CHECK-POINTS AND LAWS

Invasive species can make entry to any alien land through seaports and airports along with plant or animal products. India had a Plant Quarantine Act, which has been updated in 2003. The Government of India's efforts in the past did not yield desirable results largely because of lack of skilled manpower to detect invasive pests and weeds in seeds, plants and plant products. The new order mandated a 'Pest Risk Analysis' study for any new product entering India. The depth and extent of analysis may not be always intensive. India has only 200 technical staff to handle this vast problem area, reportedly only 7 per cent of that in China and 5 per cent of that in USA (DTE, February 29, 2004, p. 33). In United States, even with an impressive number of quarantine officers, intercepting more than 50,000 alien species a year, 'it is highly likely that at least tens and thousands more enter the country, unintercepted every year' (Mc Nelly, 2004). So one can guess the situation in the countries like India with an voluminous increase in trade under a globalised economy. The island states are specially vulnerable to invasive species as has been seen in Hawaii or in New Zealand and Australia. New Zealand has made commendable efforts to set up a federal umbrella agency with enough budget and strict regulations to detect the invasives but such efforts may not be possible in India in view of geographical locations and territorial limits. One is to be reminded that more than 90 per cent of the 115 extinct bird species were endemic to the island ecosystems.

The World Trade Organization (WTO) through several round of consultations agreed on Sanitary and Phytosanitary measures on 1st January 1995. Obviously bogey of trade barrier in the name of strict quarantine rules can be raised. But considering the procedure of 'International Plant Protection Convention' the new order of the Government of India follows such procedure. Phytosanitary certificates between the trading partners to prevent the introduction and spread of alien invasive species is one such agreed procedure.

One of the most difficult problem areas identified is introduction of alien species through release of ballast water from ships across the continent in the port of entry. International Maritime Organization stipulated methods of managing such problems by prohibiting release of ballast water within the coastal water of the country of call. The UN Convention on the Law of the

Seas also suggested states to take all measures to control intentional or accidental introduction of species, alien or new (Article 196), but in reality, the procedure is often violated in major port areas of India.

Although officially prohibited, reports from all major ports of India suggest a lack of compliance largely due to absence of effective monitoring and control by the concerned authorities. The issue of marine bio invasion has become a global concern (Pederson, 2003).

International initiative on the issue of monitoring, documenting and alarm-watch could be evidenced in the formation of IUCN's Invasive Species Specialists Group with assistance from GEF. A number of exhaustive publications especially for North America, Europe and Australia are now available (See Invasive Species Service Trace – Library of Congress) but information in invasive species in South and South East Asia remain very poor.

5.0 DISCUSSION

The richness in biodiversity is best expressed in terms of endemic species. The endemic species in many parts of the world, besides being threatened by Human activities (land use changes) is equally threatened by invasive species, introduced intentionally or accidentally. In South African Mediterranean climate 25 per cent of the species, largely woody plants are recorded as alien; of the 70 critically threatened or recently extinct taxa, 23 per cent are reportedly threatened by invading acacia, 8 per cent by pines and 2 per cent by hakeas (Hall, 1979). Detailed analysis like this has never been carried out in India. A school of researchers, however, feel that the invaders may increase biological diversity so long they do not seriously interrupt ecological system's functioning. Many of the invaders, as noted earlier may become serious pest and cause economic havoc. An estimate shows that weed invasion cause a loss of $ 37.8 million to crops in India and Arthropod invasion to crops amounts to $ 16.8 million; the figure for invasive plant pathogen related loss is put at $ 35.5 million (Pimental & Raton, 2002). The total cost of damage for India is put at $ 91.02 million as against $ 78.5 million for US. Another estimate shows that only in three developing countries in the South viz., India, South Africa and Brazil annual losses due to alien invasive species amounts to $ 138 billion per year (Pimental, et.al., 2001).

The network of protected areas is considered to be an assurance for future conservation of natural ecosystems and species. Mc Neely (2004) has cited that a number of globally important protected areas like Galapagos National Park, a World Heritage Site, Kruger National Park, South Africa's largest protected area, Everglades National Park, USA, a World Heritage Site and many others are affected by numerous invasive alien species. As such a special watch is needed to keep off the alien species from the 600 odd protected areas in India; currently, there is no reliable data on invasive alien species from the PA network in India.

The enrichment of biodiversity by exotic has been noted all over the world. But as Mooney (1988) has noted we generally know little about the effects of most invaders on the ecosystem or for the matter, the effect of most species on natural communities. Global initiatives led to Global Database on the Invasive species.

The Global Invasive Species Database was developed by the IUCN/SSC Invasive Species Specialist Group (ISSG) as part of the global initiative on invasive species led by the Global Invasive Species Programme (GISP). It provides global information on invasive alien species to agencies, resource managers, decision-makers, and interested individuals. The database focuses on invasive species that threaten biodiversity and covers all taxonomic groups from micro-organisms to animals and plants. Species information is supplied by expert contributors from around the world and includes: species' biology, ecology, native and alien range, references, contacts, links and images. The database is currently being populated with species information. (Check on *issg@auckland.ac.nz*).

Some of the invasive species like *Eichhornia* have been used for biocomposting and also for aquatic pollution control but the studies remain in pilot project scale. In Tamil Nadu a tribal group called Koravas have utilized *Lantana* material for making basket, which in turn would help for livelihood improvement programme. But such a process remains very restricted to certain geographical locations.

However, in the Indian context the knowledge on invasive species remains extremely inadequate and deserves extensive research under the initiative to conserve biological diversity in the country.

REFERENCES

Babu, Suresh, 2003. Niche Opportunity: a paradigm in invasion ecology. University of Delhi (Mimeo).

Choudhury, A. K., 1972. Controversial *Mikania* climber: a threat to forests and agriculture. *Indian. For.*, 98 (3): 178-186

Datta, A. C., 1988. Biology and Control of *Mikania*, a serious weed of tea cultivation in North East India. p. 7-16. In Studies on Weeds and their control (eds. D. N. Barthakur & A. K. Ghosh). Meghalaya Science Society, Shillong.

Dreke, J. A. et.al., 1989. Biological Invasion: a global perspective. SCOPE – ICSU Publication. Wiley, New York, 525 pp.

DTE, 2004. It's a Pogrom out there. p. 27-34. *Down to Earth*, February 29, 2004. New Delhi.

Ganeshaiah, K. N., 2003. Predicting the potential geographical distribution of sugarcane wooly aphid using GARP and DIVA – GIS. *Curr. Sci*, 85 (11): 1526

GBA, 1995. Global Biodiversity Assessment. UNEP, Cambridge University Press.

Ghosh, A. K., 1978. A brief review of works on biological control of weeds in India region. p. 82-87. *In* Studies on Weeds and their control (eds. D. N. Barthakur & A. K. Ghosh). Meghalaya Science Society, Shillong.

Hall, A. V., 1979. Invasive Weeds. p. 133-147. Report 40, CSP, CSIR, ? South Africa. 166 pp.

Lapointe, Eugene, 2006. Alien Invasions – a matter of concern for all Ecosystems. Sustainable eNews from IWMC – World conservation Trust.

Mc Neely, Jeffrey, A., 2004. Strangers in Our Midst: The Problem of Invasive Alien Species. *Environment*, 46 (6): 16-30.

Mooney, Harold, A., 1988. Lessons from Mediterranean Climate Regions. p. 157-165. *In* Biodiversity (ed. E. O. Wilson), National Academy Press, Washington D.C.

Pederson, Judith (ed.), 2003. Marine Bioinvasion: Proceedings of a Conference, January 1999 at MIT. Kluwer Academic Publishers, 143 pp.

Pimental, D., et.al., 2001. Economic and Environmental Threats of Alien Plants, Animal and Microbes Invasions. *Agriculture, Economics and Environment*, 84: 1-20.

Pimental, David and Raton, Boca, 2002. Biological Invasions: economics and environmental costs, plants, animals and microbe species, CRC Press, 369 pp.

Ramani, S., and Bhumannayar, B. S., 2003. Classical Biological Control, biodiversity and invasive pests in India (Mimeo).

Important Websites

Animal and Plant Health Inspection Service of the U.S. Department of Agriculture
Includes news, "hot issues," and a search feature.
http://www.aphis.usda.gov/

From Microbes to Mammals - Invasive Species
Articles written by the U.S. Geological Survey.
http://www.usgs.gov/invasive_species/plw

Global Invasive Species Programme
Site features news items, articles, and an extensive list of links.
http://www.gisp.org/

Institute for Biological Invasions
A very extensive site from the University of Tennessee. In addition to the information on the site itself, it includes a searchable literature database on exotic species and a wide-ranging list of links.
http://invasions.bio.utk.edu/

Invasive Plants
Produced by the United States National Arboretum. Provides general information plus links to state government departments or invasive plant workgroups and links to other major information sources.
http://www.usna.usda.gov/Gardens/invasives.html

Invasive.org
A joint project of the Bugwood Network and several government and university organizations. Its overall objective is " ... to provide an accessible and easily used archive of high quality images related to forest health and silviculture, with particular emphasis on educational applications."
http://www.invasive.org/

Invasivespecies.gov

"A gateway to Federal and State invasive species activities and programs." Includes species profiles, laws and regulations, resources, databases, etc.
http://www.invasivespecies.gov/

Invasive Species Information Node

" ... creates a central repository for information pertaining to the identification, description, management and control of invasive species." Includes species information, data and maps, and other information.
http://invasivespecies.nbii.gov/

IUCN - The World Conservation Union

http://www.iucn.org/
This organization provides a number of features of interest to scientists and the general public, including the following:
Invasive Species Specialist Group
http://www.issg.org/

Global Invasive Species Database

http://www.issg.org/database

General information on the alien invasive species issue

http://iucn.org/biodiversityday/index.html

Management of Invasive Species

AgNIC Wildlife Damage Management; provides information and links
http://lib.colostate.edu/research/agnic/invspecies

The Nature Conservancy Invasive Species Initiative

http://nature.org/initiatives/invasivespecies
Its Weed Control Methods Handbook is at
http://tncweeds.ucdavis.edu/handbook.html

Nonindigenous Aquatic Species (U.S. Geological Survey)

" ... a central repository for accurate and spatially referenced biogeographic accounts of nonindigenous aquatic species. Provided are scientific reports, online/realtime queries, spatial data sets, regional contact lists, and general information."
http://nas.er.usgs.gov/

Noxious Weeds in the US and Canada

"... a searchable database of the noxious weed lists for all U.S. states and six southern provinces of Canada. The database can be searched by plant name, state name, or by clicking on a map." Part of the INVADERS Database System from the Agricultural Research

Service, U. S. Department of Agriculture.
http://invader.dbs.umt.edu/Noxious_Weeds

SGNIS: Sea Grant Nonindigenous Species

A searchable collection of research publications and education materials produced by Sea Grant programs and other research institutions; includes links.
http://www.sgnis.org/

Southeast Exotic Pest Plant Council

http://www.se-eppc.org/
Its Invasive Plant Manual is available at http://www.se-eppc.org/manual

Union of Concerned Scientists

Includes policy and legislative information, publications to download, and other resources.
http://www.ucsusa.org/global_environment/invasive_species

Weeds Gone Wild (Alien Plant Working Group of the Plant Conservation Alliance)This extensive National Park Service website features fact sheets, plant lists, publications, information links, etc.
http://www.nps.gov/plants/alien

CHAPTER XIII

GLOBAL WARMING

1.0 GLOBAL WARMING : DIMENSIONS

What will happen with predicted temperature rise? This question has evoked serious interest and scientists across the world are studying the various possible impacts of such a scenario. The studies undertaken under International Geosphere – Biosphere Programme (IGBP) on Global Change include studies on Land-Atmosphere, Atmosphere-Ocean, Land-Ocean relationship and the changes over time and space.

With increase in computing power during 1980's, it has been possible to develop general circulation models (GCM); climate GCMs provide a global, 3-D simulation of the dynamic behaviour of the atmosphere, based on physical properties of its constituents. Simulation models studies first show that if present trends of emission of GHGs continue, average global temperature may increase between $1.5^0C – 4.5^0C$ or upto 5^0C with associated changes in the rainfall and accelerated rise in global sea level (Global Change: Reducing Uncertainties, IGBP, 1992).

Key players, Ocean, Land, Atmosphere are now being probed deep. Oceans play a vital role in controlling distribution of today's climate on planet earth. 'Oceans store and distribute vast quantities of heat but also play a significant role in absorption of atmospheric carbon dioxide by both biological and physical processes' (Pugh and Gaur, 1992).

Scientists have estimated that 'of the annual release of 5-6 billion tonnes of carbon to the atmosphere by human activities, only 3 billions remain as increased atmospheric carbon dioxide'. The oceans absorb remaining 3 billion tonnes, except for some absorbed by the green cover on land.

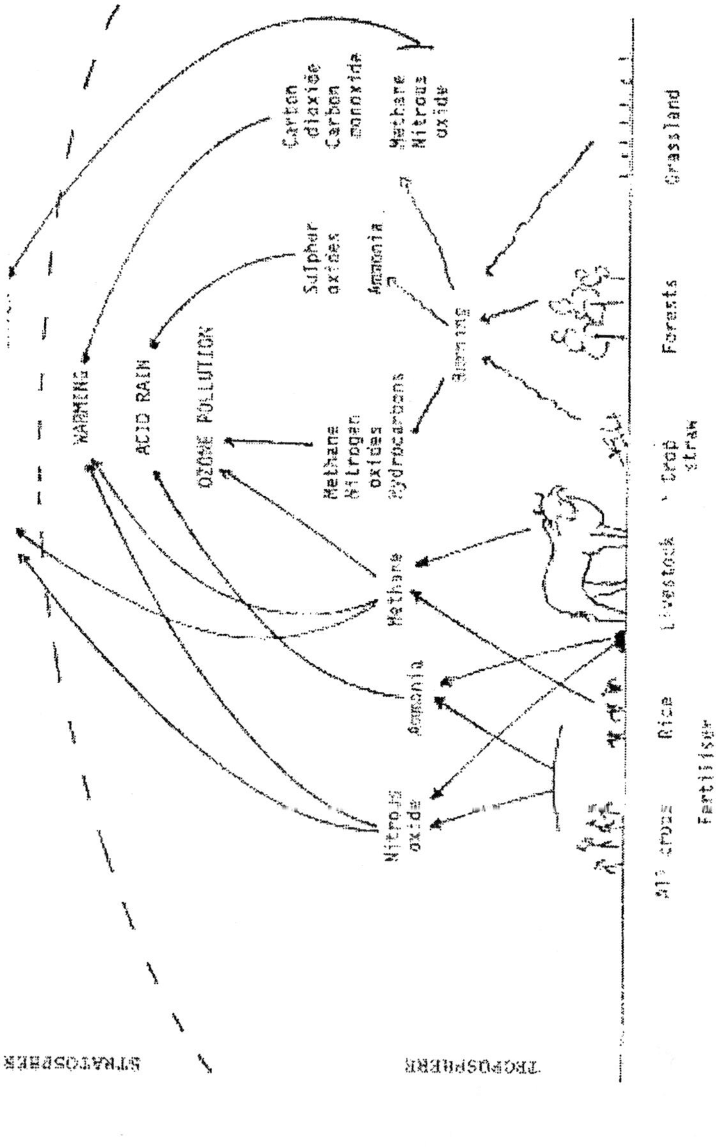

Fig-13.1: Schematic diagram of the Green House Gases originating from Cropland and Livestock

Forests on land is developed by growth of trees which in turn is dependent on the process of photosynthesis; rate of photosynthesis can be calculated by measuring the uptake of carbon dioxide from the air by the plants. Its not only the forest trees but also green vegetation that use atmospheric carbon. However agriculture is also blamed for production of another GHG, Methane, specially from irrigated rice-paddy field. Both forests and agriculture are land based system connected with the carbon and methane cycle. Land use and land cover changes (from Forests to Industrial / Residential and vegetation cover to open land area) therefore are important yardsticks to measure and understand the mode of carbon release.

It may be noted that while carbon dioxide constitute 60% of the GHG's, methane contributes about 18%. However Global Warming Potential (GWP) for methane is now estimated at 36 in comparison to carbon dioxide, but life of carbon dioxide in the upper atmosphere will be ten times more than methane.

Not only Methane or Carbon monoxide or Carbon dioxide but Nitrous oxide from fertilizer use and Ammonia, which can all contribute towards atmospheric emissions.

A schematic diagram of the Green House Gases originating from cropland and livestock is given in (Fig-13.1).

Atmosphere, surrounding the earth, the only planet harbouring living forms, is the ultimate destination for all land based emissions, other than the part absorbed by the oceans. The natural constituents of Atmosphere provided the right ambience for survival of all living forms. Its basically the chemical composition of the Atmosphere which is to undergo changes and scientists have been conducting research to understand the interactions between atmospheric chemical composition and the driving forces of biosphere and climate.

It is obvious that land environment has already been extensively altered due to human activities; doubling of human population in next fifty years will intensify such pressure.

2.0 THE DAY AFTER TOMORROW

Its not only the day after tomorrow we are all concerned, but in terms of time frame, the clock is ticking away. What will happen if temperature as predicted by scientists really rises steadily and stealthily?

Consumption of fossil fuels and the rapid expansion of industrial and domestic markets have drastically altered the atmospheric concentration of gases. When J. C. Farmen, team leader of British Antarctica Survey warned the world that 'the total ozone column is reduced by 50% (and) in a layer about 12 km thick, centered around 17 km altitude, the destruction exceeds 95%' - the world got alarmed. 'The depletion of stratospheric ozone leads to an increase in ultra violet radiation reaching the surface of the earth ..... A decrease in ozone forces a warmer climate' (Farmer, 1992). But its not only depletion of ozone, its the accumulation of GHGs, specially Carbon, Methane and others which the scientists are most worried about. Because global warming is really happening. Now.

The scientists have undertaken extensive researches on the climate change scenarios (temperature, precipitation etc.), keeping in mind the socio-economic aspects in terms of future changes in land use, economy, population etc.

The impacts of global warming can be broadly considered under

- Sea level
- Water Resources
- Agriculture
- Forestry
- Health
- Industry

and as such it covers almost all vital sectors of Human Development.

2.1 Sea Level Rise

More than 25% people in today's world live within 50 km of the coastline. India as a maritime nation has 7500 km of coastline stretching over 9 maritime states and Islands of Andaman and Nicobar and Lakshadweep. Livelihood of coastal people largely depends on coastal fishery, coastal agriculture, coastal industries (Refineries, Fertilizer plants, Petrochemical industry, Steel industries,

Power plants), coastal tourism, ports and harbour etc. If global warming leads to rise in sea level, life and livelihood of 250 million people will be at stake. Salt water ingression will destroy the coastal agriculture and vegetation and also salinize the freshwater aquifer below the ground. Habitats of fishers will be at stake as also their normal catches. Billions of rupees of investment in coastal industries will face danger. Low lying coastal areas will submerge – e.g., Sundarbans in West Bengal with 100 islands will face a tremendous crisis of survival as most of the islands are located below the sea level and are protected of embankments.

Scientists in India are studying the possible impacts in two ways – one studying the likely increase in mean sea level and other studying the frequency and intensity of coastal surges and storms.

Analyzing past tide gauge data, scientists at National Institute of Oceanography, Goa estimates 1 mm / year rise but net sea level rise can only be estimated once the data on vertical land movement is available. Scientists at the School of Oceanography, Jadavpur University, Kolkata however, have already provided a more alarming picture.

In the Sundarbans, it has been observed that the annual mean sea level has risen steadily in recent times – between 1985 to 1998 to a minimum of 4 cm rise in total 14 years and after all necessary corrections, scientists conclusively arrived at an annual sea level rise of 3.14 mm per year and 'by 2050 the compound sea level elevation will become close to 1 meter'! (Hazra et.al., 2002). The scientists' team also have shown a strong correlation between sea level rise and erosion and accretion (new land forms) taking data from Islands of Sagar, Chaksardwip, Namkhana, Mousuni and Jambudwip in Sundarbans. As a consequence, both mangrove forest area and the human population will be affected; mangrove forests providing nursery to the food fish, crabs, shrimps will diminish and at least 5,00,000 people in Sundarbans coast may be rendered homeless.

Besides the Sundarbans, A & N Islands, Lakshadweep group of islands and Gulf of Kutcchh in Gujarat are said to be the most vulnerable areas. A study showed that with 1 meter sea level rise, 50% of the total Indian coastal population will be affected. Of the estimated loss, land loss has been ranked first which in turn would lead to a large scale inward migration of population away from the coastal zone. Scientists have calculated the economic cost of sea level rise under different subject areas viz., Land loss, Loss of households, Population displacement, Dislocation of jetties, Loss of Road Networks, Cost of raising river bunds etc. And the estimated total stands at Rs. 184,766.54 crores (Asthana, 1993). The figures in terms of present day economic valuation will obviously increase.

Table- 13.1: The Economic Costs of Sea level rise 1m Scenario for India

		Land loss (in km^2)	@ Rs. / sq. Mt.	Rs. Crores
Rural		5,544.5	100	55,445.00
Urban		218.55	800 **	97,295.80
Total		**5,763.05**		**152,740.80**
	Households		@Rs/- per house	
Rural		501,029	60,000	3,006.17
Urban		255,683	250,000	6,392.07
Total		**756,712**		**9,398.24**
	Population displacement		@Rs/- per person	
Rural		5,100,000	10,000	5,100.00
Urban		2,000,000	10,000	2,000.00
Total		7,100,000		7,100.00
	Dislocation of Jetty / Ore loading points		@Rs/ Zone/ points	
Zone / Points		26/87	5,000,000	37.9
			2,000,000	
	Loss of Road network (2 m average width)		@Rs/ Sq. Mt.	
Unmetalled		3,379.65 km	300	202.78
Metalled		865.22 km	500	86.52
Total		**4,244.87**		**289.30**
	Rising River Bunds and flood protected structures by 2m height and 1m width		@Rs / cubic mt.	
Length (km)		379.43	200	15.17
Salinity prevention cost				222.00
Grant total				**184,766.54**

** For Mumbai Rs. 10,000 per sq.mt [source: Asthana, 1993]

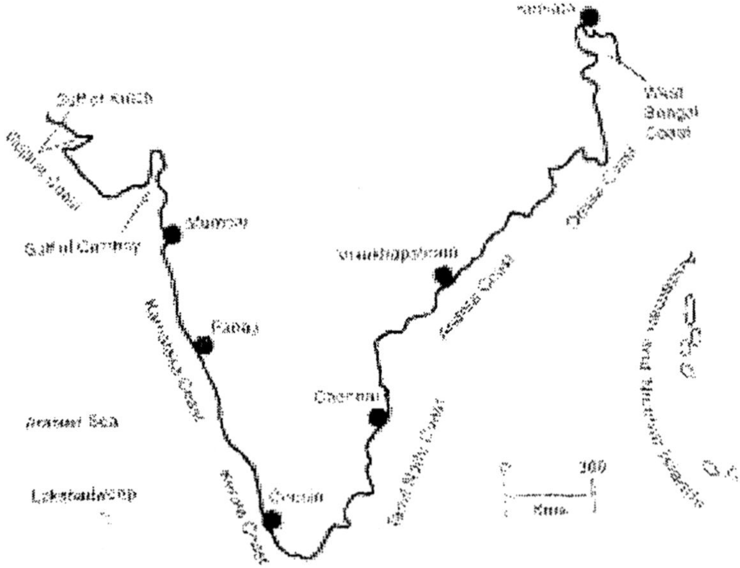

Fig 13.2: Major Maritime Cities along Indian Coast

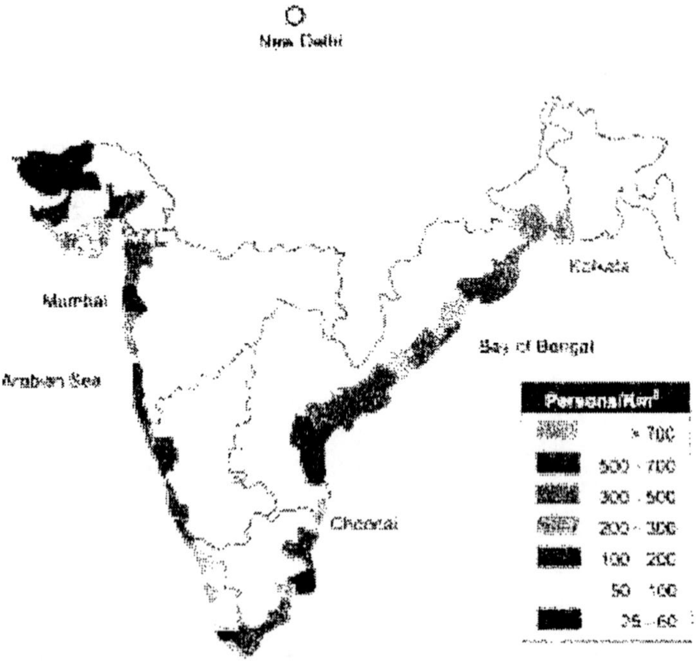

Fig 13.3: Population density along coastline of India

2.2 Water Resources

To find out impact of global warming on Water Resources, scientists at the Indian Institute of Tropical Meteorology (IITM), Pune have undertaken a study. This was based on assessment of possible impact of climate change on the country's three major river basins viz., the Ganga, the Krishna and the Godavari. 'Summer – monsoon rainfall normally contribute 70-90% of rainfall in these basins'.

Analysis of daily observed rainfall data from about 1000 stations throughout the country for the period 1901-1995 was undertaken to assess the pattern. Using computer modeling, the scientists at IITM predict that

- Hydrological cycle will be more intense with higher rainfall and increased droughts
- Extreme rainfall and rainfall intensity in all three river basins will occur by the end of 21st Century (Table-13.2)
- Number of rainy days may decrease in the western parts of Ganga basin but will increase over most parts of Godavari and Krishna basins.

Table- 13.2: Possible Changes in Rainfall and Annual Flow Pattern of Rivers due to Global Warming

River Basin	Baseline (1961-1990)		Future (2071-2100)	
	Annual Rainfall (cm)	Annual Flow (km^3)	Annual Rainfall (cm)	Annual Flow (km^3)
Krishna	91	60	112	67
Godavari	166	98	201	116
Ganga	134	482	150	543

[Source: Key Sheet No. 5, *www.defra.gov.uk/environment/climatechange*]

A focus on the Glaciers of Himalaya under Global Warming showed a positive significant retreating phenomenon (Box-1).

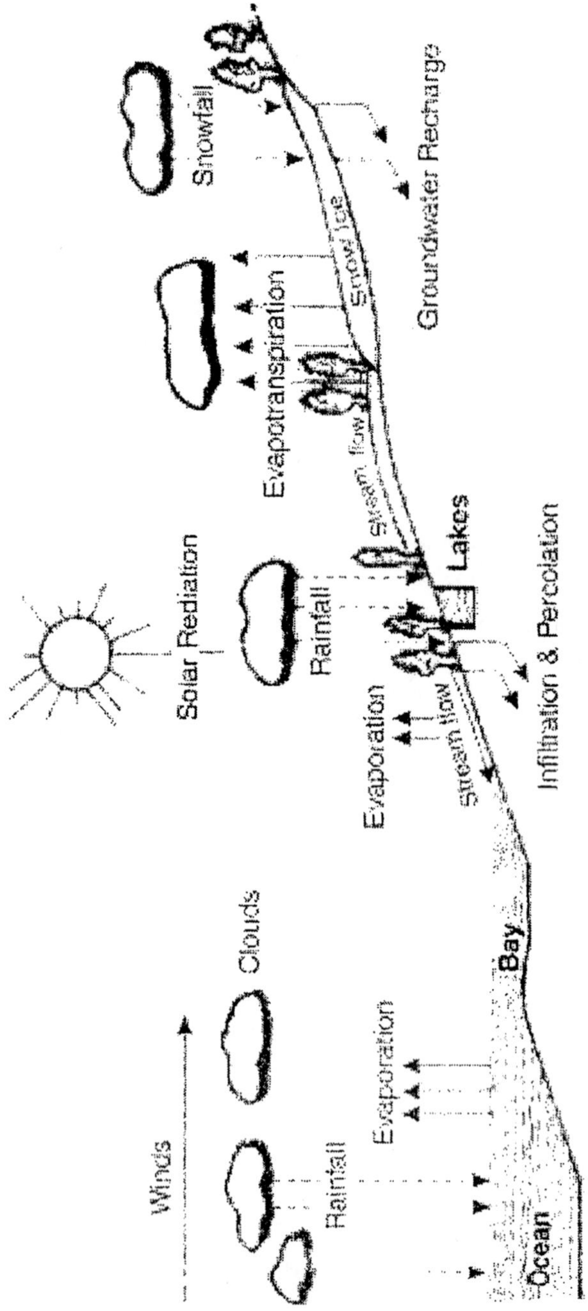

Fig-13.4: Water Cycle in Nature

Box-1: Glaciers in the Himalaya

The Indian Himalaya contains about 5000 glaciers covering nearly 38,000 square kilometers and mostly belonging Indus and Ganges river basins. Their sizes vary from less than a km^2 to 72 km long Siachen Glacier. The distribution of glaciers are higher in the North West than in North East. Alarming recession rates of glaciers during recent years has initiated a debate on possible effect of global warming in these glaciers. Currently Himalayan glaciers are retreating with an average rate of 18 to 20 m / year. Such a process is expected to be accelerated with increasing temperature and unpredictable snowfall.

[Source: Hima – Paryavaran, 16(2): 9 – 10, 2004]

It has to be remembered that 'majority of precipitation during winter monsoon occurs in the form of snowfall over the peaks of Himalaya. Himalayan glaciers cover about three million hectares or 17% of the mountain area and form the largest body of the ice outside the polar caps. The 15,000 odd Himalayan glaciers form a unique reservoir of fresh water resources, which support perennial rivers like Ganga, Brahamputra and Indus and their tributaries'. These rivers are the lifeline of millions of people in India and neighbouring countries of Nepal, Pakistan, and Bangladesh. Recent studies are showing a positive receding of these vital Himalaya glaciers, which can lead to both flooding in near future and water scarcity in distant future.

Such a scenario will call for better management of river basin, flood control and drought. Selection of crops for each planting season is to be reviewed in terms of change in water resources. Wastewater treatment and sewerage system will need to be redesigned to cope up with the overflow. All such changes will need lot of good planning and financial investment.

2.3 Agriculture

Agriculture provides the food security of a country. With increasing human population, demand for food has trebled in post independence India. To cope up with the demand of food, the green revolution was launched in 1960's; use of high yielding seeds, chemical fertilizers and pesticides and irrigation water became the major requirement for green revolution.

India's agro-climatic zone based crops depend to a great extent on monsoon rain, annual mean temperature and precipitation. The impacts of global warming, it is feared may threaten country's food security and threaten the livelihood activities, on which rural population largely depends.

Climate change can affect the plant growth, scientists say, both positively and negatively. The types of crops that are typically grown in certain area like *Jowar* and *Bajra* in Western India and Rice paddy in West Bengal can be affected due to changing rainfall and temperature; since amount of solar radiation is important for plant growth, variation in radiation pattern will have critical impacts. Higher temperature may also help breeding of tropical insect pests faster.

Scientists at the Indian Agricultural Research Institute, New Delhi is studying various range of crop growth models to assess the potential climate change impacts on Wheat and Rice – the major food crops in India and also other crops like Maize and Sorghum. Such models using variable temperature, levels of carbon dioxide, precipitation and solar radiation are expected to present likely scenario in future.

The results of the studies so far conducted show that increases in temperature by about 2^{0}C can reduce potential grain yields in most places. Rain fed crops, as opposed to irrigated crops are likely to be affected in terms of loss of yields. Similarly rice yield in likely to be reduced with an increase of 2^0 - 4^{0}C temperature. Eastern Indian states, the model studies showed, can be most impacted by increased temperature and decreased radiation.

The extent of possible loss in production has been quantified in some selected crops (Sinha and Swaminathan, 1991). It is predicted that 1^{0}C increase in temperature will reduce the duration of wheat crop by one week; one week reduction in crop duration may lead to loss of yield by 400-500 kg/ hectare. Although possible loss in rice production due to changing temperature has not been quantified, adverse changes in precipitation, will seriously affect rice production, 'since nearly 60% of rice area is rain-fed', (M. S. Swaminathan, 2002).

Conway (1992) in a paper titled 'Agriculture: culprit and victim' had shown the other relation between the process of global

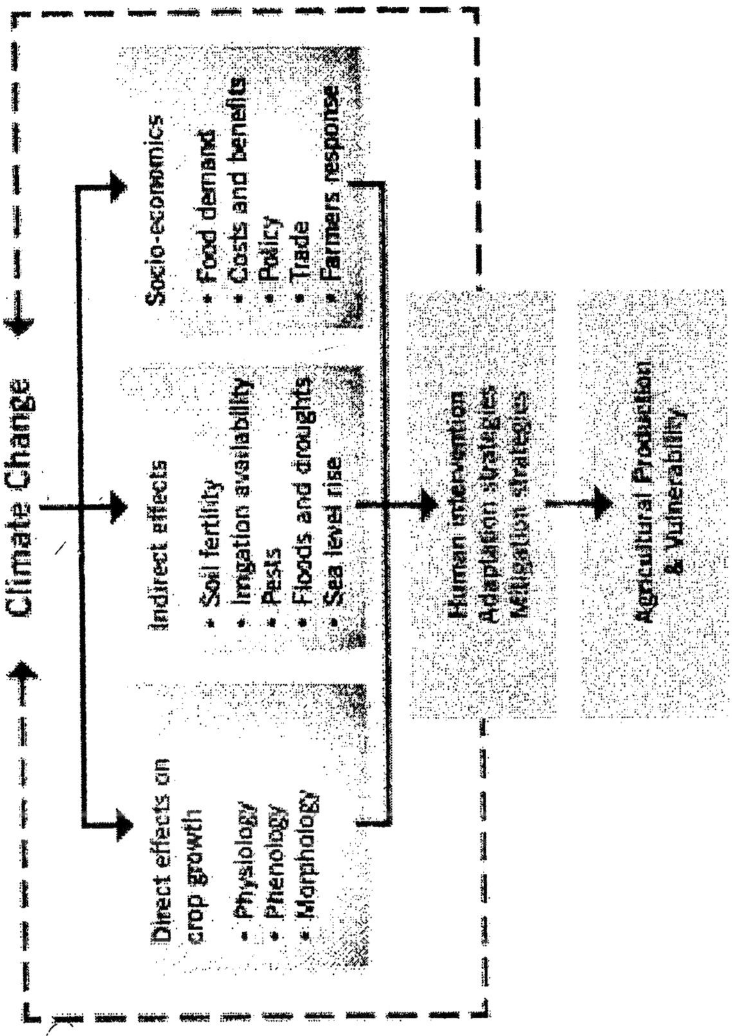

Fig-13.5: Assessment of Vulnerability of Agriculture to Climate Change

warming and agriculture. Agriculture is known to be a contributor to methane, contributing about 45% of the total emissions. 'Methane is procured by specialized bacteria in environments that are free of oxygen and during burning of biomass'. This anaerobic condition occurs naturally in swamp marshes and rice paddy field.

A preliminary estimate from Indian rice paddy has been calculated between 3 to 9 million tonnes per year. (Parshare et.al., 1991). Indian large livestock population also produces methane from activity of anaerobic bacteria, breaking down organic matter. Individual cattle, buffaloes, camels can produce 35-55 kg of methane/ year. India having the largest cattle population in the world therefore contributes substantial methane emission. Methane has a significantly higher global warming potential but lower life span than Carbon.

Besides Methane, Nitrous Oxide, resulting from use of nitrogen fertilizer, also contributes to the global warming. Nitrous oxide is produced naturally by the action of bacteria, specially in tropical soil.

While all these gaseous emission contribute to the global warming but as shown earlier, the impact of agriculture made it one of the significant victims of process of climate change.

The global warming therefore may demand alternative cropping pattern for each agro climatic zone. This is a challenging exercise and scientists are yet to come up with the alternative plan to deal with the situation.

2.4 Forestry

'Forests' for long time was taken as a resource base for timbers. Timbers for industrial use gained ground in international trade and forestry practices have widely been modified by modern human society. Natural forests gave away to man-made forests in many parts; diversity of natural forests changed into commercially valuable forest stands – often mono-cultural. Gradually, the concept has changed. That forests not only are important sources of timber but act as habitat for major land based biodiversity is now well known. Forests provide services by way of soil conservation, recharging water resources, absorbing carbon dioxide and releasing oxygen. Forests help in precipitation and rainfall. All these knowledge have now put a new pricing for forests along with vast resources called Non-Timber Forest Produces (NTFP). People in and around the forests depend heavily for their livelihood on forest resources.

In India, forests occupy 22% of the land area. Management of forests no longer depends on Forest Department, the mode of Joint Forest Management (JFM) i.e., jointly with the local community has gained ground since 1988.

It is well known that climate exerts profound control over the process of exchange of CO_2 and O_2 in photosynthesis of plants. Besides solar radiation and ambient carbon dioxide, moisture, temperature and soil nutrition are critical variables. Carbon dioxide uptake generally increases with increasing temperature to a level but then tends to decline. Carbon dioxide in the fully matured trees remains limited as against growing forest stand. Low temperature, chilling and frost – act as stress factor reducing the rates of photosynthesis.

As temperature increases, activities in the plants enhances till an optimum level but at very high temperature plants may perish. Temperature influences the activity of plant enzymes, that triggers photosynthesis and respiration, affects solubility of CO_2 in the plant cells, transpiration and ability to absorb water and minerals from the soil.

While global warming is directly linked with carbon built up in the atmosphere, a higher induced increase in respiration could cause a significant additional release of CO_2 reinforcing the very build up that initiates the warming process (Readers may remember that in plants photosynthesis help to absorb CO_2 but respiration as in all other system release CO_2). Forest trees can effectively act as carbon sink when the process of absorption is higher than the net release.

Foresters opined that changes in climate may alter the success of regeneration and growth of plants as also the mortality rate of tree species.

Scientists at the Indian Institute of Sciences, Bangalore in collaboration with Forest Survey of India, Dehra Doon and the Indian Institute of Tropical Meteorology, Pune undertook a recent study to determine how global warming and climate change will impact different forest biomes (broad categories of forest dependent on climatic condition). Baseline data from 11 different biomes were gathered on past climate, vegetation and the soil. Using computer based climate models, projections are made for 2070 and 2100. Comparing the present climate scenario with future probable climate (and CO_2 concentration) computation was carried out to study the impacts on biomes. As a general rule, regions showing a greater variance between the two are assumed to be more vulnerable to climate change.

The results show that country's dominant forest types namely Moist Savanna (32.5%) and Dry Savanna (33%) will be changed to an extent that Tropical Dry Forest (37.2%) and Tropical Seasonal Forest (28.4%) become dominant. Likewise vegetation in the dry region will change from shrub land to woodland. In colder region, Temperate Conifer and Boreal Forests will decrease and Temperate Deciduous and Evergreen Forest will increase. Approximately 90% of India's forest cover shows a possible change. Such an event may lead to large scale forest die back and loss of biodiversity. While these will have serious adverse impacts, 'Net Primary Productivity' (NPP) – a measure of amount of vegetation matter produced, is likely to increase in the majority of locations due to carbon fertilization'.

Implication of such large scale change have socio economic ramifications. People dependent on forests may find it difficult to harness NTFP – used for fuel, fodder, medicinals, gums, resins and other purposes.

A special focus on forest changes and biodiversity impact can now be given. Animals, from soil microbes and micro arthropods (insects, mites, spiders) to large mammals are all adapted to not only a typical habitat but also to climatic conditions. Migratory animals have regular cycle of winter migration and return migration at the end of winter. With global warming, animals will tend to migrate from their home range, the time of seasonal migration may also alter and boundary shifting in the protected areas will be almost certainly be necessary. Species sensitive to higher temperature may be severely affected and 'some species that are currently classified as critically endangered' will become extinct without being able to adapt to new climate regime.

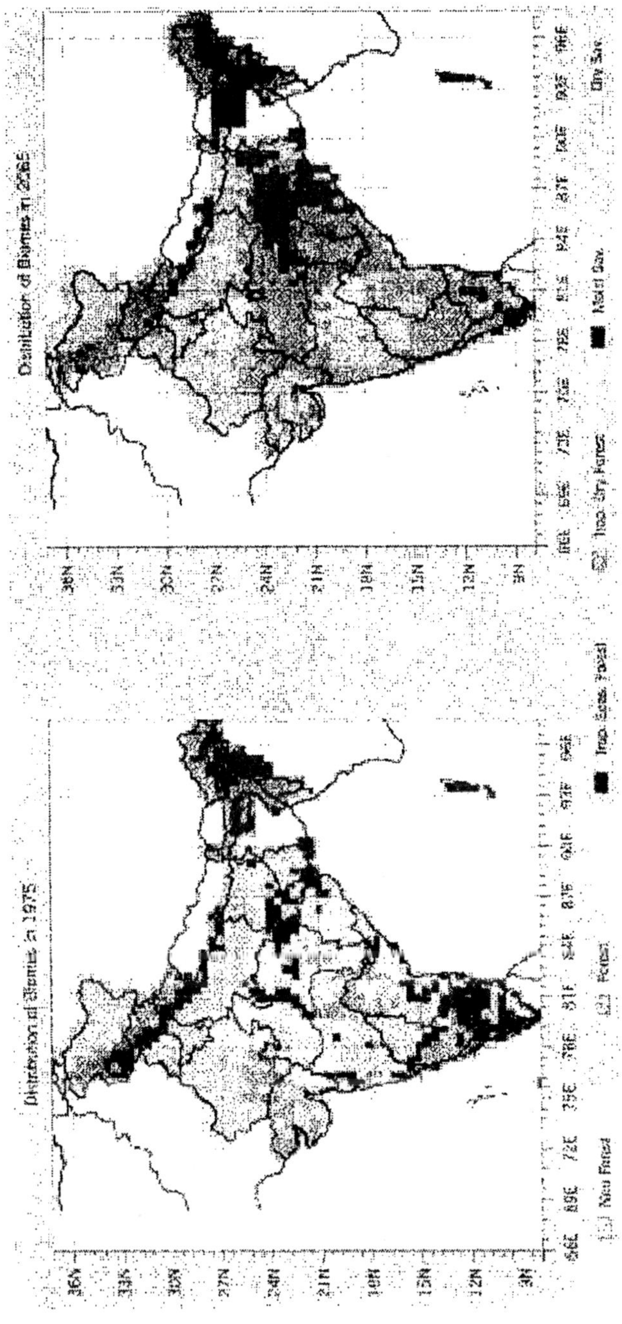

Fig-13.6: Impact on Forest Biomes

Table -13.3: Biome Types in India – Current and Projected

Biome type	Number of grids (Current)	Number of grids (Projected)	% Current	% Projected	Change
Boreal evergreen	999	292	2.81	0.82	-1.99
Boreal / temperate mixed	0	859	0.00	2.41	2.41
Temperate conifer	1074	488	3.02	1.37	-1.65
Temperate deciduous	2063	2314	5.80	6.50	0.71
Temperate evergreen/warm mixed	1895	3539	5.32	9.94	4.62
Tropical seasonal forest	711	10098	2.00	28.37	26.37
Tropical dry forest	2719	13237	7.64	37.19	29.55
Moist savanna	11577	198	32.52	0.56	-31.97
Dry savanna	11771	1544	33.07	4.34	-28.73
Xeric woodland	624	2708	1.75	7.61	5.85
Xeric shrub-land	2010	254	5.65	0.71	-4.93

[Source: Key Sheet No. 7, *www.defra.gov.uk/environment/climatechange*]

Scientists at National Institute of Oceanography, Goa predict that with mean sea level rise intertidal habitats will have direct effects with increased influence of marine process. The mangrove forest can not expand towards landward side and many such area 'may totally disappear as a result of permanent submergence of roots and pneumatophores' (breathing roots) (Jagtap and Komarpani, 2003).

2.5 HEALTH

The health of Human Population depends on reasonably stable climate, protection of ultra violet (U-V) radiation (which can cause amongst others, skin cancer), adequate supply of food and freshwater and survival of biological resources. Till middle of 20th Century, all these were taken for granted or only localized water crisis, flood, famine etc., occurred. Often such crisis were met by importing goods, migrating population or innovating new strategy.

Global warming, scientists believe can change not localized but global life support system. U-V radiation and increasing incidences of skin cancer can be a wide spread concern. Increasing temperature has shown increased incidence of malaria at higher altitudinal range. A model study showed that 3^{0}C warming by 2100 would increase malaria by 100 times. One of the major direct impacts may be heat wave related death. Already unusual temperature has caused death in new areas across the globe. Higher temperature helps in faster breeding of vector (insects, mites) and can invariably lead to an increase in vector borne diseases.

The loss of biodiversity will make the world poorer in terms of biological resource and range of their diversity. Each living organism, variety, land races carry a characteristic set of gene (hereditary material) and larger the gene pool better is the chance of selection for hybridization. Besides, many of the species have never been tested about their potentiality of use in future either as food, fodder or medicine. There are numerous instances of amazing discovery of useful characters in unsuspecting plant, animals or microbial organisms. Each of these depends on specific climatic condition for survival. Drastic change may lead to a serious crisis of existence.

The relation between air pollution, global warming and heath has also been studied by the scientist. Photochemical pollutant and aero-biological agents (spores, moulds etc.) may cause serious respiratory diseases with the rising temperature.

The changing agricultural pattern can make food security of people a critical issue; in turn, lack of nutrition in such areas would make the people vulnerable to diseases; chances of person to person infections are likely to increase. Besides Malaria, dengue fever zone may see a new high of incidences.

Forecasting the health impacts of global warming is a complex and an unusual challenge. Direct and indirect impacts may be too many due to changes in the surrounding of the habitats human society has lived and adapted to over millennia.

The range of global warming will invariably determine the extent of health impacts. The possible major types of impacts of climate change and ozone depletion in stratospheric ozone (stratosphere is a part of upper atmosphere) is shown in the diagram below (Fig-7).

A special mention can be made to the study conducted by the National Physical Laboratory (NPL) in India on Malaria – current scenario and possible change of scenario after predicted climate change. It shows that malaria incidences will be more in terms of number of months, depending on the arrival of monsoon.

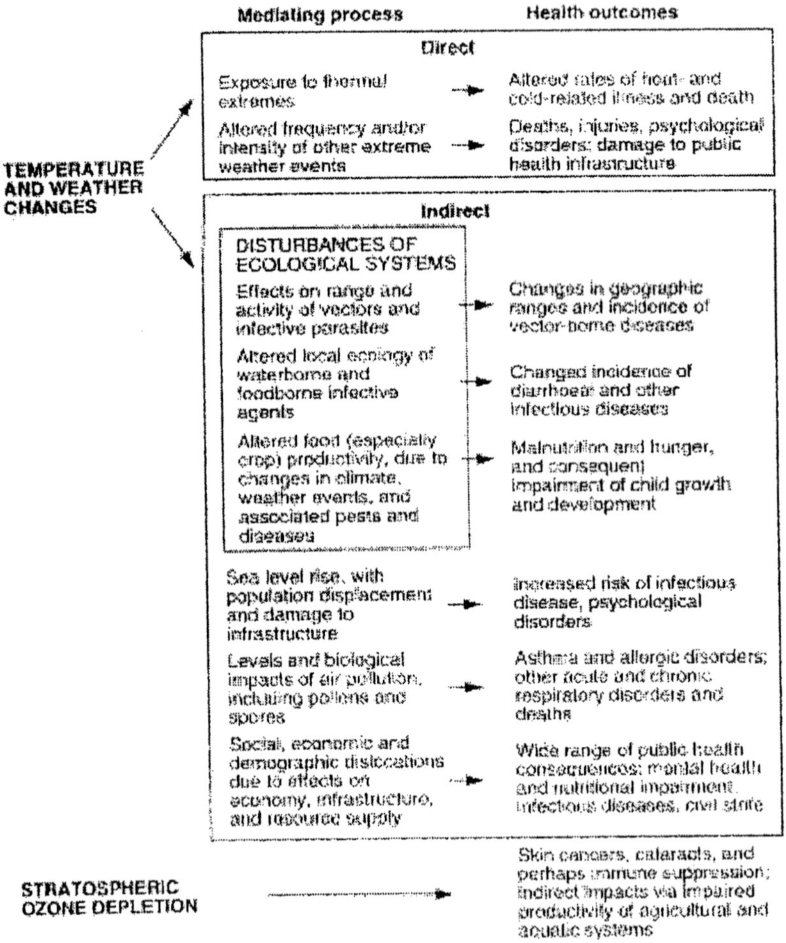

Fig - 13.7: Possible major types of impact of climate change and stratospheric ozone depletion on human health

2.6 INDUSTRY

Industrial development in India has been accelerated with opening up of market, new trade initiatives and investment. India produces all essential industrial products from paper pins to Aero planes; small, medium and large scale industries are coming over even in hitherto unattended regions. Foreign investment has contributed to the second industrial resurgence - era of India.

But global warming can have serious adverse effects on these economic investments. Infrastructure for industries may be major target areas of global warming, like energy and transportation. Scientists believe that Physical infrastructure is generally vulnerable to climate, specially with increased rainfall, rising sea level and extreme heat etc. The Indian Institute of Management, Ahmedabad (IIMA) carried out research to investigate the possible impacts of climate change and global warming on industry, energy and transport. The following diagram (Fig-8) shows the impacts of climate change on infrastructure.

Researchers specially focused on Transportation by studying Konkan Railways Corporation Limited (KRCL) – a coastal railways of 720 km long along Western Ghats of India. IIMA studies show the possibility of direct adverse effects on strength of building material and indirectly the need for enlarging air conditioning system; rainfall increase may lead to erosion of land endangering the track; sea level rise can also cause soil erosion, water logging and flooding which in turn will demand additional investment to maintain the railway track.

The present power generation scenario, IIMA studies show, project nine folds increase from 96 GW to 912 GW between 1995 – 2100. Climate change may necessitate 1.5% more power generation. Unpredictable rainfall may increase irrigation need and power will be needed to run pumps for abstraction of groundwater. Temperature rise will increase demand for cooling, which will in turn demand for oil. As India largely depends on coal-based power, more demand for power will lead to more use of coal and increased carbon emission. Thus a vicious cycle may be foreseen.

All these forecasting indicate that 'Global Warming' can have a range of impacts endangering the very foundation of development.

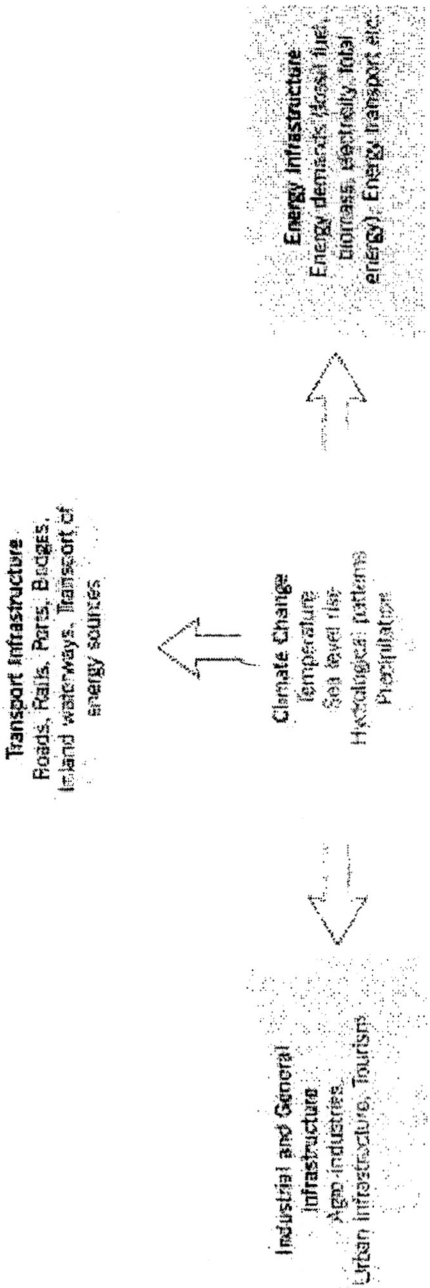

Fig-13.8: Impact of Climate Change on Industry, Energy and Transport

3.0 COMBATING GLOBAL WARMING

But while the Human Society is held responsible for the impending dangers of global warming, it's the responsibility of the same to solve the problem. Such an effort has been initiated way back in 1992. At the United Nations, the first ever U N Framework Convention on Climate Change (UNFCC) was adopted. The Convention, now 15 years old, has 189 signatories till date (March 2007). The Convention established a process of periodic meetings called Conference of the Parties (COP). The 12th Conference of the Parties of UNFCC held at Nairobi; Kenya in November 2006 was the first ever such conference in the South of Sahara, Africa. Scientists believe Africa will have one of most critical impacts of global warming (Harmeting and Dossou, 2007).

UNFCC acknowledged that change in Earth's climate and its advance efforts are a common concern for humankind and it decided to protect the climate system for present and future generations. It must be mentioned that earlier UN took initiative to have a 'Convention for the Protection of Ozone Layer' in 1985, commonly called Vienna Convention and a Protocol on Substances that Deplete the Ozone Layer in 1987, known as Montreal Protocol. It provides a roadmap for reduction of use of such substances, which affect the ozone layer including a deadline fixed for change of technology. India is a party to UNFCC, Vienna Convention and Montreal Protocol.

UNFCC has 26 Articles and Article 4 has enlisted the future commitments by the countries, who have agreed to abide by the roadmap. It has specially listed the developed countries under two lists – Annexure I and Annexure II and clearly defined the responsibilities of the countries enlisted. It calls upon all developed countries under Annexure I (Table - 13.10) to adopt policies and take measures for mitigation of climate change by limiting man-made emissions of GHG's. It called for returning to emission level of 1990 for carbon dioxide and other green house gases. The developed country parties and other developed parties included in Annexure II (Table - 13.4) are to provide new and additional financial resources to meet the full costs for adoption of steps to curb carbon emissions incurred by developing country parties.

Table -13.4: List of Annexure I and II Countries

Annexure I	Annexure II
1	2
Australia	Australia
Austria	Austria
Belarus	Belgium
Belgium	Canada
Canada	Denmark
Czechoslovakia	European Economic Community
Denmark	Finland
European Economic Country	France
Estonia	Germany
Finland	Greece
France	Iceland
Germany	Ireland
Greece	Italy
Hungary	Japan
Iceland	Luxembourg
Ireland	Netherlands
Italy	New Zealand
Japan	Norway
Latvia	Portugal
Lithuania	Spain
Luxembourg	Sweden
Netherlands	Switzerland

1	2
New Zealand	Turkey
Norway and Northern Ireland	United Kingdom of great Britain
Poland	United State of America
Portugal	
Romania	
Russian Federation	
Spain	
Sweden	
Switzerland	
Turkey	
Ukraine	
United Kingdom of Great Britain and Northern Ireland	
United States of America	

[Source: United Nations, 1992]

Developed countries have shown some positive actions in capturing carbon dioxide and developing new technologies to make profitable use of excess carbon (see Box 2)

Box 2: Capturing Carbon Dioxide

THE CENTRE for Innovation in Carbon Capture & Storage (CICCS), which has recently opened at Nottingham University, England, has been established to develop novel technologies to trap the 'greenhouse gas' carbon dioxide and store it permanently and safely, so that it is not released into the atmosphere.

The aim is for the centre to facilitate interdisciplinary activity, taking groundbreaking ideas from basic science and developing them into new products, processes and services. It will also address issues of public acceptability. Within the centre a new generation of potential academic, industrial and government leaders in CCS will be trained with a broad and interdisciplinary set of skills suitable for their future careers in industry, research or government.

One of the technologies that the centre will work on uses a natural process in conjunction with silicate-based rocks such as serpentine. This is found in large enough quantities, and in the right places, to store all the carbon dioxide produced by the combustion of the entire world's known fossil fuel reserves. The carbon dioxide extracted from burning coal is put into a reactor with the rocks and through a chemical reaction the serpentine binds the carbon dioxide to itself, locking it in permanently.

Compared with other proposed processes for carbon storage (such as burial under the sea), once the carbon dioxide is locked inside the rock by the CICCS process, it is combined forever and cannot go back to its previous state. This is one of the method of reduction of carbon dioxide that has so far been the most controversial issue in carbon storage.

Fossil fuel power plants could take advantage of the new process by adding a reactor to their emissions treatment system. Allowing carbon dioxide to be turned into a useful building material. The centre's ultimate goal will be to sign collaborative agreements with power and construction companies to move forward with the commercialization of this technology.

[Source: *Britain Today*, 2008 – Vol I, Page 36-37]

Efforts have also been made to join hands together for collective action. One such is International Carbon Action Partnership (ICAP), launched by European Union. ICAP will facilitate such global solution by rigorously monitoring, reporting and verifying emissions and working to determine reliable sources appropriate for inclusion in a globally linked programme. (see Box 3)

Box 3: Fighting Global Warming

A COALITION of European Union countries recently announced the formation of the International Carbon Action Partnership (ICAP) to fight global warming. ICAP will provide an International forum in which governments and public authorities adopting mandatory greenhouse gas emissions cap and trade systems will share experiences and best practices on the design of emissions trading schemes. This cooperation will ensure that the programmes are more compatible and are able to work together as the foundation of a global carbon market. Such a market will boost the demand of low-carbon products and services promote innovation and allow cost-effective reductions so as to enable swift and ambitious reductions in global warming emissions. The groundbreaking international and interregional agreement was signed by the UK, Portugal, Germany, France, Ireland, Italy, Spain, the Netherlands, Greece, the European Commissions, the USA and Canadian members of the Western Climate Initiative, northeastern USA members of the Regional Greenhouse Gas Initiative, as well as New Zealand and Norway, who joined on behalf of their emissions trading programme.Global warming is a problem that requires a global solution. ICAP will facilitate such global solution by rigorously monitoring, reporting and verifying emissions and working to determine reliable sources appropriate for inclusion in a globally linked programme. It will encourage common approaches; create a clear price incentive to innovate, develop and use clean technologies; encourage private investors to choose low carbon projects and technologies as well as provide flexible compliance mechanisms that ensure reliable reductions.

[Source: *Britain Today*, 2008 – Vol I, Page 33]

Since the developed countries have contributed maximum GHG's, UNFCC put the responsibilities largely on such countries for funding, transfer of technology to meet specific need and concerns of developing countries specially to the :

i. Small Island Countries

ii. Countries with low lying areas

iii. Countries with arid and semi-arid areas, forested areas and areas liable to forest decay

iv. Countries with areas prone to natural disasters

v. Countries with areas liable to drought and desertification

vi. Countries with areas of high carbon atmospheric pollution

vii. Countries with areas with fragile ecosystems

viii. Countries whose economies are highly dependent on income generated from consumption of fossil fuels, and

ix. Land locked and transit countries.

Africa may be an area of high risk because of "its economic dependence on natural resources and agriculture. Water and rainfall matter a great deal, and droughts and floods means major disaster, immediately threatening livelihood". Before proceeding further, one may look at the UN Millennium Development Goal (MDG), which has pledged to reduce the poverty in the World by 50% by 2012. Six years after the MDG's were first adopted, scientists feel, global warming has finally risen to the top of international agenda. Since global warming is likely to affect food production, energy supply and transport and also healthcare and other social sectors, it has a strong linkage with the issue of alleviation of human poverty. And Africa is most challenged under such a scenario.

In the UNFCC arrangement, two international funds have been set up, one called the 'Special Climate Change Fund' (SCCF) and the other 'Least Developed Countries Fund' (LDCF). Both funds depend on voluntary contributions from the developed industrialized nations. Third fund called 'Adaptation Fund' (AF) has been set up later to help implement the time frame and the programme (of cutting down GHG's) provided by the Kyoto Protocol. It is to help tangible adaptation measures in developing countries. The fund will be generated by taxing levy on the trade on carbon (in issuing emission certificate). Carbon trading is now accepted as a method under Clean Development Mechanism (CDM); Industrial units in developed countries emitting more carbon than permitted, could under the system help developing countries reduce their emission and get credit for the same. By 2012 the AF is expected to collect \$ 270 - \$ 600 million. This will be used to change to clean technology but the concerned people feel that more funds will be necessary for the developing countries to adopt clean technologies (e.g., change from conventional energy to renewable energy).

To address specific problems related to each of the causes – i.e., emission of GHG's, new strategies are now being suggested or devised as follows:

Methane emission

To combat methane emission, which has more global warming potentials, changes in agricultural practices are being suggested. Developed countries especially USA blamed India and China for methane emission.

Rice cultivation

To change rice cultivation, suggestions are being made viz.,

i. Irrigation practice with more area under frequent drainage (instead of flooding)

ii. Adopting varieties with low methane emission potential and high yield

iii. Substituting inorganic fertilizers by organic fertilizers, which is cable of reducing emission by 50%

iv. Changing water management system promoting mid season aeration by short term drainage

v. Applying nitrate fertilizers in crops with aerobic conditions to reduce emission of nitrous oxide.

Auto emission

Likewise, to control auto emission, a major contribution to carbon load, a national auto fuel policy can be launched. The one in India includes:

a. Fixing mass emission norms for petrol vehicle and diesel vehicles

b. Use of catalytic converters

c. Setting up mandatory norms at manufacturing stage for all petrol or diesel vehicles, in 4 wheelers

d. Change into unleaded gasoline

e. Reduce benzene content in gasoline to 1%, sulphur content to 0.05%

f. Use hybrid fuel e.g., gasoline blended with ethanol

g. Use Compressed Natural Gas (CNG) and Liquefied Petroleum Gas (LPG) to control emission.

In general fuel quality standard and vehicular technology need drastic changes.

Water crisis

To resolve possible *water crisis*, several suggestions have been forwarded viz., Integrated Management, Management at Basin Level, Dam and Management, and Recycle and Reuse of Wastewater. Emphasis has also been given on Rainwater Harvesting either for recharging of groundwater or as a supplement to other water uses. Special focus on the rational utilization of water for irrigation has been given in the National Water Policy, 2000.

Forestry

In the *forestry* sector, afforestation, reforestation and slowing deforestation can be the keywords to combat global warming. It is calculated that such measures can reabsorb 60-87 Gt (giga tonnes) of carbon on 700 million hectares between 1990-2050.

Coastal zone

To address the issue of *coastal zone*, India has enacted "Coastal Zone Regulations" to control and regulate developmental activities in 9 maritime states and island groups. Prohibiting structures within High Tide Line and 500m on the landward side, is one of the restrictive guidelines under CRZ Notifications, 1991.

Industries

Under new fiscal policies, *Industries* are pursued to promote low emission technologies and cut energy cost. Environment Management Plan (EMPs) and Environmental Audits are being more effectively implemented in the Industry sector. Clean technologies are promoted in 3 million Small Scale Industries. Refrigeration and Air Conditioning industries have already taken steps to cut down CFC's (which contribute to ozone depletion and global warming).

Energy

In the pursuit to advocate *Renewable Energy* use, new ventures in solar, wind, biomass, hydel power are launched. A separate ministry of Non Conventional Energy (MNES) provides extra institutional support. A new National Rural Energy Policy, 2005 further provides guideline for sustainable use of biomass energy besides others. Nearly 40 million trees have been planted over the coalfields and environmental planning with afforestation programme has become mandatory along with mining plan.

In the *Oil and Natural Gas* sector, efficiency of boilers and furnaces is being improved with waste heat recovery system at major installations.

The recent report of IPCC has given the possible impact with each one degree centigrade rise in temperature in a scale of 1-6^{0}C (Table – 13.5).

Table - 13.5: Hot Pursuit

Increase in temperature (^{0}C)	Impacts
1^{0}C	• Shrinking of glaciers threaten water for 50 million • Modest rise in cereal yields in temperate regions • 300,000 die from malaria, malnutrition and other climate related diseases each year
2^{0}C	• 15-40% of species face extinction, including Polar Bears • Potential for Greenland ice sheet to start melting irreversibly, committing world to 7m sea level rise
3^{0}C	• 1-3 billion more suffer water crisis • 150-550 million more at risk of hunger • 1-3 million more die of malnutrition • Amazon slump starts • Risk of abrupt changes to monsoon

4^0C	• Crop yield falls 15-35% in Africa • Up to 80 million more exposed to malaria in Africa • Loss of half Arctic tundra. Loss of more species. More deaths from malnutrition
5^0C	• Glacier melting hits a quarter of Chinese and millions in India • Ocean acidity kills ecosystems and fish • Rising sea threatens small islands, Florida coast, New York, London, Tokyo
6^0C	• There is a 10 per cent chance. Catastrophe for all life on Earth

[Source: IPCC Report Summary in *The Telegraph*, 12th February 2007]

In summary, while India and the World is facing a great danger of Global warming, concreted efforts are being made to control the process. Model studies were conducted to evaluate socio-economic scenarios with possible changes in 4 different situations viz., (i) Government controlled economic growth, (ii) Policy shift towards sustainable development, (iii) Private capital participation in social sector, (iv) Use of environmentally sustainable technologies have shown 4 different profile; the parameters used were Population, Gross Domestic Product (GDP), Food grain demand, Food grain productivity, Water demand, Demand for electricity at the household level, Wood demand, Fire wood demand. The Energy and Resource Institute's studies projected the findings. The results show that within 30 years (1990-2020 and 2020-2050) there may be significant changes, in each of the projected scenario but it becomes obvious that government / private investment would promote economic and industrial growth, a policy shift to sustainable development and global concern will put more priorities to the social and environmental values (Table 13.6-13.11).

Table-13.6: Population Projections (millions)

	Scenario I	Scenario II	Scenario III	Scenario IV
1990s	846	846	846	846
2020s	1354	1012	1291	1228
2050s	1888	1646	1572	1298

Table-13.7: GDP Projections (Rs. Crore)

	Scenario I	Scenario II	Scenario III	Scenario IV
1990s	886933	886933	886933	886933
2020s	5094093	3833274	8924904	5866501
2050s	14298068	9304363	33426607	19027395

[one crore is 10000000 rupees. One lakh is 100000 rupees]

Table-13.8: Projections of Food grain Demand

	Scenario I	Scenario II	Scenario III	Scenario IV
Demand for food grains (million tonnes)				
1990s	176	176	176	176
2020s	295	220	304	267
2050s	450	392	449	309
Required food grain productivity (tonnes / hectare)				
1990s	1.7	1.7	1.7	1.7
2020s	2.3	1.7	2.4	2.0
2050s	3.5	3.0	3.5	2.2

Table-13.9: Projected Total Demand for Water (billion cubic meters)

	Scenario I	Scenario II	Scenario III	Scenario IV
1990s	629	629	629	629
2020s	2609	2444	3421	2258
2050s	3682	3576	4341	4846

Table-13.10: Demand for Electricity by Urban Households (million tonnes of oil equivalent)

	Scenario I	Scenario II	Scenario III	Scenario IV
1990s	3.9	3.9	3.9	3.9
2020s	23.4	17.6	33.8	22.5
2050s	69.4	49.0	69.6	53.3

Table-13.11: Demand for Wood

	Scenario I	Scenario II	Scenario III	Scenario IV
Demand for Industrial round wood (million cubic meters)				
1990s	26.8	26.8	26.8	26.8
2020s	33.4	31.4	39.3	34.6
2050s	36.5	33.7	44.9	38.6
Demand for firewood (million tonnes)				
1990s	122.2	122.2	122.2	122.2
2020s	203.5	152.0	200.5	199.9
2050s	270.0	235.5	247.3	222.8

[Source: Key Sheet No. 3, *www.defra.gov.uk/environment/climatechange*]

4.0 LOOKING FORWARD

The Kyoto Protocol; binding on the developed countries as biggest carbon emitter, is to end by 2012. A review in 2012 was the topic of discussion in the recently held Conference of the Parties (CoP) meet at Bali, Indonesia, in 2007 (Box 4).

Box 4: Bali Summit 2007: Historic Breakthrough

A HISTORIC AGREEMENT on a road map for achieving a global climate deal by the end of 2009 was reached at the recently held conference in Bali. After two weeks of intense negotiations, developed and developing countries signed up to the agreement. This is the first time ever that all countries of the world got together to negotiate on a climate treaty to take the world beyond 2012.

The Bali Roadmap sets out a clear and comprehensive agenda for negotiations and a timetable ending in 2009. It also links back to the work of the Intergovernmental Panel on Climate Change (IPCC). In particular, all countries will adopt a shared vision for long – term cooperative action, including a long-term global goal.

The talks also resulted in breakthroughs on deforestation, adaptation, technology, finance and the carbon markets. On technology, there was agreement on an ambitious work programme covering both mitigation and adaptation that provides the basis to accelerate and scale – up investment in technology beyond that of the carbon market.

The agreement in Bali will pave the way for incentives to reduce emissions from tropical deforestation. This includes both deforestation and more gradual damage short of total deforestation. It also sets the rules for projects to reduce deforestation; including allowing projects to be piloted to a common UN-approved standard, with the aim of the experience gained feeding into a future climate framework. The guidance makes it clear that national baselines should be developed and gives primacy to national-level approaches.

A decision was reached on the governance of the Adaptation Fund, which supports adaptation in developing countries funded by a levy on the Clean Development Mechanism (CDM). As for carbon markets, it was decided to abolish registration fees and levies on CDM projects in the least developed countries, improve the way the CDM and its Board functions and approve the use of non-renewable biomass CDM. In other words, projects such as small cooking stoves would now be possible through the CDM.

[Source: *Britain Today*, 2008 – Vol I, Page 34-35]

India and China targeted by developed countries especially USA was put to dock. The logic was that in a global scenario per-capita emission does not work. India and China are, overall big contributors to the emissions. India and China argued that cutting down carbon based gases will slow the progress of these emerging economies. But the most recent Intergovernmental Panel on Climate Change (IPCC) Report, 2007, claims that the loss of GDP, if the countries undertake cuts, will not be more than 3% of GDP. A look at the Tables 12 & 13 showing cumulative emissions from 10 countries and also emissions in 2003 alone would show the logic of current arguments and debate.

Table 13.12

CUMULATIVE EMISSIONS 1950 - 2003

Rank	Country	Mt CO_2	World Total	Tons CO_2 Per Person	Ranking*
1	USA	230,200.00	26.37%	791.6	2
2	EU (25)	187,773.90	21.51%	411.6	24
3	China	83,515.60	9.57%	64.8	91
4	Russian Federation	81,779.50	9.37%	565.6	11
5	Germany	49,946.20	5.72%	605.1	8
6	Japan	41,057.30	4.70%	321.8	35
7	UK	31,415.70	3.60%	527.3	12
8	India	22,098.00	2.53%	20.8	124
9	Ukraine	21,722.20	2.49%	454.3	18
10	France	19,854.90	2.27%	330.8	33

* per capita emissions

Table 13.13 : Emissions in 2003 alone

Country	Tons CO2 Per Person	Rank*
Qatar	44.5	1
Kuwait	26	2
UAE	23.9	3
Luxembourg	22.8	4
Bahrain	22.6	5
USA	19.9	6
Trinidad & Tobago	19.5	7
Brunei*	18.3	8
Canada	17.2	9
Australia	17.2	10
India	11	120

* per capita emissions

With risk of global warming, India, as already stated could face crisis in food security. A FAO Report in 2007, stated India could lose as much as 125 million tonnes of its rain fed cereal production. Rain fed agriculture in marginal areas in semi-arid and sub humid regions is mostly at risk. This is in contrast to the predicted crop yield production at higher latitudes for global average temperature increase of up to 3° Celsius, depending on the crop and their disease, beyond that.

The recent Human Development Report (HDR, 2007) of United Nations warned that unless Developed Countries cut emission at least by 80% by 2050, progressively from 20-30% in 2030, earth can not be saved from a 'Complete Environment Catastrophe'. However, it also stipulated that fast growing economies like India and China must share the burden by cutting down 20% carbon emission. If both developed countries and emerging economies act, the cost could be only 1.6% of global GDP by 2030. HRD report states that while developed countries with 15%

world population is responsible for 50% share of Carbon emission, developing countries would be worst affected by possible displacement of 330 million people due to flood and malarial infection of 220-400 million people; 262 million people were affected by natural calamities between 2000-2004.

The year 2007 saw frantic activities at all level to understand, not the reality of global warming, but how to offset the process of crisis. After the 13th meeting of COP to the UN Frame Convention on Climate Change (UNFCCC) held in Bali, in 2007, R K Pachauri, the Head of IPCC (who brought out Fourth Assessment Report in 2007) urged in a media article "Lets go beyond Bali" (*Times of India*, December 24, 2007). After outlining the threat, the author stated that a multilateral agreement is planned under UNFCCC and the Kyoto Protocol, to be put in place by the end of 2009. Importance of the 4th IPCC Report is re-emphasized in the findings that "Warming of the climate system is 'unequivocal' and most of the observed increase in global temperature is due to human induced Greenhouse Gas emission. Evidence of an increase in intensity and frequency of heat waves, droughts, floods and extreme precipitation events in later part of 20th century are recorded. Melting of glaciers and increasing scarcity of freshwater add worries. Inspite of serious debates on specificities, a compromise has been arrived on actions of mitigation, including "Measurable, reportable and variable nationally appropriate mitigation actions by developing country parties in the context of sustainable development, supported by technology and enabled by financing and capacity building".

In view of imminent and multiple threats facing India due to global warming, urgent action is called for. Following the path of developed country, will prove to be flawed. India can set the example for other developing countries in devising a truly sustainable development pathway. Increasing energy efficiency, Fuel switching, changing industrial process, utilizing municipal solid waste may be some of the open ended options. Emphasis on renewable energy specially, biomass based, still remains one of the chosen preference.

Tracing carbon foot print is the newest game in the scenario. Every country has a carbon foot print but how such prints can be curbed? Technology transfer and funding support remain two major demands from developing countries. How much will be archived is yet to be seen.

REFERENCES

Asthana, V., 1993. Reports on impacts of sea level rise on the islands and costs of India. Report submitted to Ministry of Environment and Forests, Government of India, New Delhi.

Bradley, Raymond, S., 2000. 1000 years of climate change. *Global Change News Letter*, 44: 5-6.

Conway, Gordon, K., 1992. Agriculture: Culprit and Victim. p. 100-114. *In* Global Climate Change: Science, impacts and responses (eds. Bob Frame et.al.). The British Council, New Delhi.

Farman, J. C., 1992. The effect of Ozone depletion on climate. p. 19-25. *In* Global Climate Change: Science, impacts and responses (eds. Bob Frame et.al.). The British Council, New Delhi.

Harmeling, Sven and Bosson, Krystel, 2007. Coping with global warming. D + C, No. 3(2007), 103-105.

Hazra, Sugato, Ghosh, Tuhin, Dasgupta, Rahashree and Sen, Gautam, 2002. Sea level rise and associated changes in the Sundarbans. *Curr. Sci*, 68 (9-12): 309-321.

IGBP, 1992. Global Change: Reducing Uncertainties. 1-40.

IGBP, 2001. Global Change and the earth system: A planet under pressure. IGBP Series 4, 1-32.

Jagtap, T. G., and Komarpani, D. S., 2003. Evaluation of Mangrove Ecosystem of India for assessing its vulnerability to projected climate change. p 39-52. *In* "Assessment of Climate Change in India and Mitigation Policies" (eds. S. K. Dash and Prakesh Rao), WWF-India, New Delhi.

Jenkins, G. J., 1992. The IPCC scientific assessment of climate change: an overview. p. 33-41. *In* Global Climate Change: Science, impacts and responses (eds. Bob Frame et.al.). The British Council, New Delhi.

Prashar, D. C., Rai, J., Gupta, Prabhat, K., and Singh, N., 1991. Parameters affecting methane emission from paddy fields. *Ind. J. Radio and Space Physics*, 20: 12-17.

Pugh, David and Gaur, Vinod, K., 1992. The Critical role of the oceans in climate change. p. 41 55. *In* Global Climate Change: Science, impacts and responses (eds. Bob Frame et.al.). The British Council, New Delhi.

Sinha, S. K., and Swaminathan, M. S., 1991. Deforestation, Climate Change and Sustainabl Nutrition Security: A Case Study of India. *Climate change*, 19: 201-209. Kluwer Academic Publications, The Netherlands.

CHAPTER XIV

ENERGY, ENVIRONMENT AND DEVELOPMENT

1.0 THE SURGING ECONOMY

India is steadfastly showing an impressive rate of growth in Gross Domestic Product (GDP). Of the so called BRIC countries, India's closest competitor is China; while in 2005 GDP growth was 2.9% in Brazil, 6.4% in Russia, 01% in India, China scored 10.4%; on 2006, the respective figures changed (Table 14.1). India is chasing China with 9.4% growth in GDP. This economic growth is dependent in Energy Consumption. However, all economic growth potential is influenced by capital, labour force participation rate, productivity growth and above all by population growth.

Table 14.1: Real GDP growth in emerging markets (in %)

	2005	2006
Brazil	2.9	3.7
Russia	6.4	6.7
India	9	9.4
China	10.4	11.1

[Source: RBI Deputy governor Rakesh Mohan's valedictory address to bankers on August 3, 2007]

In a globalized economy, economic growth in India and China is said to be contributing to the real GDP growth in the World. The projected 4.1 GDP growth is considered mostly due to non-OECD countries. Rapid pace of development in these countries makes it possible to change the pattern of past 25 years to a better scenario in next 25 years.

Economic growth and energy needs are linked by energy intensity. In developed countries, energy demand is said to be trailing behind economic growth. Globally, energy intensity may decline at least by 2.3 % /year till the end of 2030. As such energy security is to be ensured to make the path of economic growth uninterrupted.

2.0 COAL FACTOR

The coal reserves of India are still the lifeline of major energy production. At a growth rate of 5 % in domestic production, currently extractable local resources are predicted to be exhausted in about 40 years, says Prof. Kirit Parikh, leading Expert Committee on the Integrated Energy Policy, constituted by the Government of India. Worse is the fact that India's proven coal reserve will run out in 80 years. Major proportion of coal extracted goes to the power sector, accounting for nearly 77% (Table 14.2). However, BP Statistical Review of World Energy provides a different picture on India's coal scenario. It states that at the current rate of production, India will use 92,445 million tonnes of proven coal reserves in 207 years. BP Report compares India's coal reserves and use with that of US, Russia, China and the World (Table 14.3). In any case the Government of India's decision to add nine Ultra Mega Power Plant (UMPP) with capacity of 4000 MW each, the demand on coal will increasingly grow. The current plan is to increase generation to 212 GW from 122 GW by 2012. Whether BP Report had taken account of this programme is not clear.

Table 14.2 : Coal India Despatches*

Year	Total Despatches	Despatches to power sector	% for power sector
2000-01	272	207	76.1
2001-02	281	217	77.2
2002-03	288	220	76.4
2003-04	303	234	77.2
2004-05	321	249	77.6
2006-06	333	256	76.9
GDP growth			

* In million tonnes; [Source: Coal India Limited]

TABLE 14.3 : PROVEN COAL RESERVES (2006- END IN MILLION TONNES)

Country	Anthracite & bituminous	Sub-bituminous and lignite	Total	Share of total %	R/P ratio
US	111,338	135,305	246,643	27.1	234
RUSSIA	49,088	107,922	157,1010	17.3	* *
CHINA	62,200	52,300	114,500	12.6	48
INDIA	90;085	2,360	92,445	10.2	207
AUSTRALIA	38,600	39,900	78,500	8.6	210
WORLD	478,771	430,293	909,064	100	147

** More than 500 years

What are reserves to production (RP) ratio: If the reserves remaining at the end of the year are divided by production in that year, the result is the length of time in years that those remaining reserves would last if production were to continue at that rate.

[Source: Statistical Review of World Energy, June 2007]

In any case coal is a finite resource and increasing dependency on coal can never be a model foı sustainable development at the current pace of its utilisation.

The future crisis in coal based energy production apart, the issue of global warming and climate change makes it compulsive to look for clean energy, curb carbon emission and contribute to the global cause. While developed countries have to share the major burden of curbing carbon emission, developing economics can no longer be playing safe by arguing that per capita emission still remains very low. It is the cumulative emission from a country thats counts now. Recently concluded meet of the Conference of the Parties (CoP) to the 'UN Framework Convention on Climate Change' (FCCC) clearly indicated need for immediate relook in 2009 on the second phase of commitment beyond 2012, under Kyoto Protocol. India has to come up with a positive action oriented plan for carbon reduction while keeping development pathway open.

3.0 THE RENEWABLES

India for long being dependent on coal based conventional energy, however, thought of possibility of use of other energy sources or non-conventional in the past. Ministry of Non-Conventional Energy (MNES) bears testimony to such a concept. However, non-conventional energy owes its production to the sources of renewable energy viz. Solar, biomass, water, air etc. Technology enabled the human society to harness electricity from such natural resources, which may promise an inexhaustible resource base.

3.1 Harnessing Water

Of the renewables, India has a long history of giving priority to the "Hydel Power", as a part of Multipurpose River Valley Projects. Hydel power is clean, non-polluting and with proven water resource in varied topographical scenario can surely offer enormous scope for power generation. Hydel power demand dams and reservoir (as a pre requisite condition) to be built. Such dams and reservoirs can lead to loss virgin tract of tropical forests with yet to be documented biodiversity; human population is affected due to inundation and consequent eviction of people. While clean power is most welcome, can it be at ether enormous environmental and / or social cost? One of the earliest environmental movements in India centred around hydel power project proposed in the Silent Valley of Kerala. Conservation of Tropical Rain Forest and Wild Biodiversity finally won over – the project was shifted!

India's hydel power potentiality is estimated at 84,000 MW at 60 % load factor. The current installed capacity is stated to 30,955 MW (in October 2007). The expert committee on energy, led by Prof. Kirit Parikh estimated that "an installed capacity of 150,000 MW, including some 15,000 MW of mini hydel plants (size less than 25 MW) may be justified with available potential hydroelectric energy". Of all the states in India, Arunachal Pradesh in North-East can take the place of pride with projected 25,000 MW of hydel power to be generated under 50,000 MW new capacity programme. Himachal and Uttaranchal in North- West also occupy significant positions with regard to hydel power generation.

Region wise break-up shows Eastern region has nearly 2500 MW installed capacity of hydel power out of total installed capacity of 16,225.28 MW representing little over 15 % of hydel power, dependency. In the West, the figure stands at 6,918.83 MW out of total installed capacity of 37,137.36 MW representing 18.6 % dependency. The North shows highest dependency on hydel power at 37.8 % with installed hydel capacity of 12,360.38 MW out of a total of 35,475.65 MW. The southern region stands second with nearly 30 % dependency on hydel pcwer (Table 14.4).

Table 14.4 : India's Hydel Power

Region	Hydel Installed Capacity	Total Installed Capacity	Percent
East	2,496.53	16,225.28	15.4
West	6,918.83	37,137.36	18.6
North	12,360	37,475.65	34.8
South	10,947.71	36,809.32	29.9

Currently, India ranks 8th in the global order with regard to dependency on hydel power at national level (at 14.3 %) when compared with other players, Norway is believed to top the list with 98.9 % level of dependency. China is now the largest producer (in terms of trillion watt hours) with 397 TWH and accounts for 13.3 % of global energy generation. Performance of certain other countries can be seen from the following table (Table 14.5).

Table 14.5:Hydel Power Generation /Dependency of Selected Countries

Norway	98.9
Brazil	83.7
Venezuela	73.9
Canada	57.9
Sweden	46.0
Russia	18.3
China	15.9
India	14.3

It is apparent that fast developing BRIC countries (see Table 14.1) show a combined hydel power dependency factor of 33%. This figure can very well be used against the accusing figure on carbon emission from such nations; however, it doesn't justify further carbon dependency but calls for more dependency on renewables.

Hydel Power has other questionable implication. A National Environmental Engineering Institute Report in 2006 showed serious impact of emerging hydel power project on the fisheries and biodiversity in Beas and Sutlej rivers in Himachal Pradesh. HP fisheries department even claimed compensation of Rs.2.5 crores from NTPC for this loss.

On the other hand "Small Hydel Projects" may play a safer role. Negative impacts of these projects are minimum; operation and maintenance cost also remains affordable. However, till 2005, only 1705 MW of electricity was generated from small Hydro Projects and an additional 480 MW was projected for projects under implementation.

Table 14.6 : State-wise SHPs (up to 25 MW as on Feb 28, 2005)

S.No.	State	In Operation		Under Implementation	
		No's	Capacity (MW)	No's	Capacity (MW)
1	2	3	4	5	6
1	Andhra Pradesh	57	178.81	5	7.30
2	Arunachal Pradesh	57	33.80	55	51.87
3	Assam	3	2.11	8	51.00
4	Bihar	5	45.90	9	14.00
5	Chattisgarh	4	13.50	1	5.50
6	Goa	1	0.05	-	-
7	Gujarat	2	7.00	-	-
8	Haryana	5	62.70	-	-

1	2	3	4	5	6
9	Himachal Pradesh	50	108.08	8	52.75
10	Jammu & Kashmir	29	105.24	7	10.31
11	Jharkhand	6	4.05	8	34.85
12	Karnataka	49	276.78	7	14.00
13	Kerala	14	84.62	6	60.40
14	Madhya Pradesh	8	41.16	3	24.20
15	Maharashtra	27	207.08	4	15.25
16	Manipur	8	5.45	3	2.75
17	Meghalaya	3	30.71	9	3.28
18	Mizoram	16	14.76	3	15.50
19	Nagaland	8	20.47	6	12.40
20	Orissa	6	7.30	7	40.97
21	Punjab	23	111.40	4	5.75
22	Rajasthan	10	23.85	-	-
23	Sikkim	12	35.60	5	15.20
24	Tamil Nadu	12	77.70	1	6.60
25	Tripura	3	16.01	-	-
26	Uttar Pradesh	8	21.50	1	3.60
27	Uttaranchal	75	72.45	38	26.01
28	West Bengal	20	92.26	7	5.80
29	Andaman & Nicobar Islands	1	5.25	-	-
	Total	522	1,705.59	205	479.30

[Source: Dam, River, People: March 2006]

It may be mentioned that potential for microhydel power in India has been estimated at 15,000MW. The Ministry of Non Conventional Energy (MNES) has identified 4233 potential SHP sites with a capacity to produce 10,324 MW, of the total 15,000 MW potential. A total of 15 states have announced policies for setting up SHP's through private sector participation.

China, by 2002 was operating 42,221 SHP serving 300 million people. Of these, 58% were managed by the communities, rest being controlled by either the Government or Private Sector. India, undoubtedly is lagging far behind China in small hydel project.

3.2 Solar power

India in the tropical world is endowed with solar energy, freely available for human use. Traditionally, rural people developed indigenous method of solar drying of grains and sundry other material.

It is reported that the average daily solar radiation varies between 4-7 kW per square meter in different geographical areas of the country. Sunny days vary between 250-300 and India receives about 5000 trillion kw of solar energy on an average, every year. However, contribution of solar energy to total energy scenario still remains very poor. (*Power Line* June, 2005)

Solar energy available from sunrays in a given day is more than the entire world consume in a year. Problem remains with cost of generation of solar power. Conventional solar cells cost heavily, because they are made from rather expensive silicon semiconductors – like the ones used in computer microchips. Search for alternative cheaper and affordable material has been made around the world. Recent success in, making solar cells using ultra thin films of silicon opened up new opportunity. With cheaper solar cells use, consumption is likely to increase. But solar power still has to compete with the cost of thermal , oil or natural gas generated electricity. It is reported that despite quadrupling in sale between 1997 – 2002, solar power accounted for only .04 per cent of worldwide power. (Fairley 2002)

Solar energy can be transformed into electricity through a two stage process. Using large mirror, high energy heat is generated from sunlight; the heat is channelled through conventional generator to change into electricity. This process of concentrating solar power along with solar thermal remains major option. But fastest growing solar technology is photo voltaic. The photo voltaic cells made mostly of the semiconductor silica convert sunlight directly into electricity.

Use of photovoltaic cells are widespread – from the simplest cells in watches and calculators to more complex system contributing to the power grids - uses encompass wide area. Usually 40 cells are combined in the modules. In USA, photovoltaic cost 20 to 30 cents per kilowatt hour while more updated wind technology costs 5 to 6 cents. Solar power remains more expensive globally, much more in countries like India.

One of the reasons for cost intensiveness lies in the fact that the conversion efficiency from solar energy to electric energy is in the range of 10 %, the other 90 per cent is lost through heat. But major advantage is their easy use even in remote areas. Researchers have enabled even to integrate photovoltaic technology with construction material and be built into a building rather than added in top. (Pellerin, usinfo.state.gov/gi/Archive/2005)

Solar Power in India

In a recent assessment, Ministry of New and Renewable Energy (MNRE) stated that 'owing to high initial cost of solar power plants and the cost of electricity from such plants (Rs. 15 per kWh in grid without battery as opposed to Rs.1.75 – 3.80 in small hydro), the ministry has decided to provide a generation based "fillip". The effort is to provide maximum of Rs.12 per kWh for electricity generated from solar photovoltaic (PV) and Rs.10 per kWh for electricity generated through solar – thermal power plants. Government of India stated that this subvention would be provided to project developers at a fixed rate for a span of 10 years. Each state will be allotted to install up to 10 MWp aggregate capacities under this programme. It is anticipated that additional investment of about Rs.1000 crore mostly for private sector, would be made. The grid – interactive power plants would be set up on a built, own, operate (BOO) model. MNRE expects that such initiative would help to preclude one metric tonne of carbon dioxide per year per one kW capacity built up; expectedly these plants would create employment opportunity for 25 – 40 persons directly and offer 400 indirect jobs.

The thrust on solar power can be judged from the global demand for solar technology; it was estimated US $ 2.0 billion in 2003, $ 15 billion in 2006 and likely to reach $ 100 billion in 2015.

According to the website of the Government of India (http//: mnes.nic.inj). India's position stands (as on September 30, 2007) as follows:

- Grid interactive solar power: 2.12 MW

- Decentralised energy system
  - Solar photovoltaic 110 MW (peak)
  - Solar street lighting System 61,321 no's
  - Solar Home lighting system 3,63,399 no's
  - Solar lantern 5,64,931 no's
  - Solar power plants 2.18 MW (peak)

- Solar Thermal Programme
  - Solar water heating system 2 million ^{2}M collection area
  - Solar cooker 0.7 million
  - Solar Photovoltaic pumps 7068 no's

The solar energy business in India is growing at a fast pace. International companies like Roth and Ram from Germany sees India emerging as the fourth largest generator of solar energy. Business is likely to jump 40 times within the current decade ending 2010. Manufacturing units for solar modules and components are being set up by multinational company (Conergy, Germany); photovoltaic sells based on silicon technology are going to be produced with investment of more than Rs.330 crores by another multinational (Moser Baer); likewise "Solar Semi-conductors" is going to invest $40 million to produce photovoltaic cells. Indian companies are also showing keen interest in solar power. Reliance Industries has pledged to put up a 10 MW solar power plant in backward Purulia district in West Bengal.

However, the government of India has given priority to provide solar power panels for lighting and meeting minimum needs of people through off – grid system to 80,000 villages. Major constraint remains in the cost. If the cost price of one kilowatt of power generated by solar photovoltaic continues to remain Rs.15 (in small scale operation), it will be difficult to promote. Better prospects, experts opine, could be in the use of solar thermal unit, which can reduce the cost by nearly 50 %. It is felt that the Government needs to come out with appropriate policies on subsidies, soft loans and tariff for promoting solar power in India. Changing over to non – emitting clean energy can help India earn through Carbon Emission Reduction Credits. (Solar Power Boom, In *Dams, Rivers and People*, 5 (10 – 11): 1, 8 – 11), 2007).

4.0 GOING WITH THE WIND

4.1 Indian Scenario

Wind power has been traditionally recognized in Indian mythology. The God of Wind being called 'Pavan' but the power in harnessing electricity was never utilised in the country to any significant level. India with abundant sunshine and wind resource has now embarked on both solar and wind energy programmes. The Government of India has estimated that the potential of wind power in India can be put at 45,000 MW but the estimate is restricted to 10 of the 29 states of the country. The renewable energy till May 2007 contributed seven per cent of the power generated and wind power is said to have a major share of 5.2 per cent. The growth in wind power shows a quantum jump, from only 42 MW in 1992, reached 300 MW by the beginning 21st century and shot up to 615 MW in 2003 and by 2007; it reached an installed capacity of 7113.6 MW. State wise installed capacity in Wind Power in India is given in Table 14.7 below.

Table 14.7 : Wind Power in India

State	Installed Capacity (in MW)
Andhra Pradesh	121.8
Gujarat	667.1
Karnataka	849.4
Kerala	2.0
Madhya Pradesh	57.8
Maharashtra	1484.9
Rajasthan	469.9
Tamil Nadu	3457.5
West Bengal	1.6
Others	1.6
Total	7113.6

[Source: Ministry of New and Renewable Energy, 2007]

It can be seen that Tamil Nadu, Maharashtra, Karnataka, Gujarat and Rajasthan are the major players in the field.

4.2 International Scenario

Internationally, Denmark in Europe became the first major commercial user in 1990's – of course Holland could claim to be the pioneer in installing wind mills. Denmark was followed by Germany but currently Spain, Italy, France and U.K. besides India are recognized as major market for wind power. In U.S., total installed wind power capacity was 9,149 megawatts at the beginning of 2006 with a plan for additional 3000 MW by end of the year. Use of wind power in U.S. has increased by nearly 100 per cent between 2004 and 2007, showing a marked shift in global energy policy. (Clean Energy Solution, *eJournal*, USA, Economic Perspective, 2006, 11 (2): 21). By mid 2007, total global installed capacity was projected to reach 78,720 MW for which four countries contributing 71 % are credited. These are Germany, Spain, USA and India. (Source: *Wind Energy Monthly News Magazine*)

Fig. 14.1 : Decline in Renewable Energy Costs

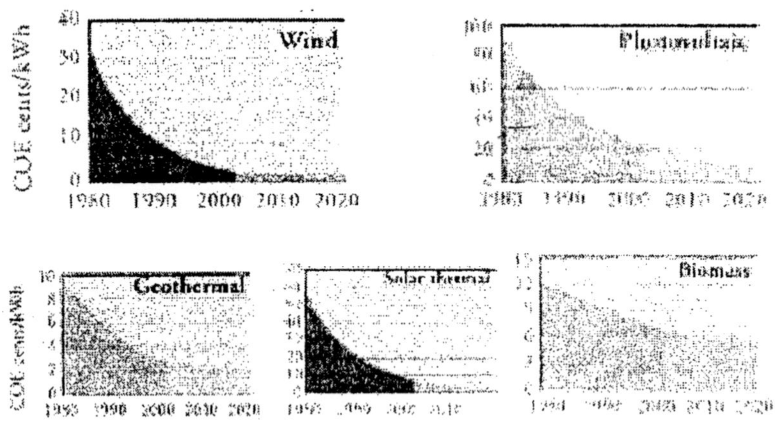

These graphs are reflections of historical cost trends, NOT precise annual historical data.

COE = Cost of Energy.

[Source: National Renewable Energy Laboratory (www.nrel.gov/analusis/docs/cost_curves_2002.ppt].

While one of the reasons may be increasing cost and crisis of global warming being linked to carbon based energy, the other reason may be the decline in renewable energy cost. Between 1980 and 2020, a projection shows significant decline especially in Wind, Solar thermal and photovoltaic energy. (Fig. 1).

4.3 Economics & Cost Benefit

The allure of Wind Power has made big corporate houses surge into the market to India. Reliance Energy, Tata Power, DLP, Suzlon, Vestas India, Indian Oil and ONGC have all become interested to play a role.

Union government of India has introduced a package of incentives including tax holidays, custom and excise duty relief etc. Investment in technology improvement for better wind turbines for higher generation efficiency have yielded results Indian Renewable Energy Development Agency (IREDA) promotes and extends financial assistance to renewable energy.

In general, an amount of Rs.5.0 – 6.0 crores per MW is required to put up a medium scale wind turbine. One of the reasons of this high cost is wind turbines currently used in India are either imported or manufactured under license. India is yet to develop the know-how to produce wind turbine – the critical point being appropriate design for blades and rotor. Once national Aerospace Laboratories, Bangalore in India develop suitable indigenous technology, 50 % of cost of import could be saved.

The concerted efforts of India to develop wind power earned India the World Wind Energy Award, 2005 for Outstanding Achievements in favouring policy for the Wind Energy. Once the important substitution is made the cost of production is likely to come down. Currently it stands at Rs.2.50 – 3.00 per unit, of course with support of fiscal benefit.

Wind Energy data shows that a single 750 kW wind turbine can help reduce 1,179 tonnes of carbon – dioxide, 6.9 tonnes of sulphur – dioxide and 4.3 tonnes of nitrogen- oxide. Clearly such figures are impressive for reducing the risk of global warming. Wind Power projects can gainfully employ 25 – 30 people per MW of installed capacity.

5.0 DOWN TO EARTH: BIOMASS

5.1 Indian scenario

The National Conservation Strategy and Policy Statement on Environment and Development, 1992 listed "Biomass" under 'Priorities and Strategies for Action', as one of the five major focal points. India is said to be a Biomass based country. It is estimated that Biomass currently contributes 14% of the total energy supply worldwide. Currently, the global capacity for generation of Biomass based power is put at 35,000 MW and India's share stands at only 2%.

India however has a potential of generating power to the tune of 19,500 MW from available fuel wood, crop residue and bagasse from sugar industry. Currently only 727 MW is being produced (Table 14.8)

Table 14.8 : Status of Biomass based Power in India

SN	State	Biomass	Cogeneration
1	Andhra Pradesh	182.2	63.0
2	Chhattisgarh	11.0	—
3	Gujarat	0.5	—
4	Haryana	4.0	2.0
5	Karnataka	36.0	116.0
6	Madhya Pradesh	1.0	—
7	Maharashtra	3.5	32.5
8	Punjab	10.0	12.0
9	Rajasthan	7.8	—
10	Tamil Nadu	34.0	138.5
11	Uttar Pradesh	—	73.0
	Total	290.0	437.0

[Source: Dams, Rivers & People, Dec 2005 – Jan 2006]

One of the most focused areas of power generation using biomass is 'stand – alone, off grid, wood based biomass' dependent power. In West Bengal, a captive energy plantation to feed 500 kW biomass power plants has set an example of NGO – Government collaboration for rural energy supply. This was achieved through meticulous planning, selection of right species, setting up nursery and public cooperation in raising plantation in Common Pool Resource land. (*Ghosh and Das 2003*).

Under Rural Energy Security Policy of Government of India, use of biomass energy (from captive energy plantation) has now been advocated. In many states viz., in West Bengal, Orissa, Madhya Pradesh and Tamil Nadu, focus on biomass based energy are evident. Use of organic solid waste, rice – husks, wood and bagasee are becoming more effective. Subsidy from the Ministry of New and Renewable Energy provides incentives for such project.

It is stated that biomass gasification technology is probably the most efficient energy conversion technology; with a combination of diesel generator set and gasifier; up to 80% diesel consumption can be reduced by using wood biomass. (*Mondal, S. K. 2008*)

Besides, supplying rural energy, biomass based energy has helped to set up captive power generation to meet energy demand in Sugar, Cement Industries, Rice Mills etc.

5.2 International Efforts

United States Congress has passed a Biomass Research and Development Act in 2000. The legislation created the National Biomass Initiative (NBI) co-chaired by officials of Department of Energy and US Department of Agriculture, to coordinate technology development. 'The NBI collaborate, internationally through the International Energy Agency, and every four years sponsors an international biomass conference to compare technologies and approaches to biomass development' (Pellerin, 2005). In May 2004, delegates from more than 100 countries attended the Second World Conference and Technology Exhibition on Biomass for Energy and Industry in Rome, Italy. This was jointly held with 13th European Conference on Biomass for Energy, Industry and Climate Protection.

Biomass use has opened the possibility of use of biomass in bio refineries. Bio refineries integrate biomass conservation processes and equipments to produce a range of fuels power and chemicals from biomass. It can take advantage of the differences in biomass components and intermediates thereby maximise the value derived from biomass. The first International Workshop on the subject was organised jointly by NBI, USA, European Commission and Canada was held in July 2005. It is believed by NBI that the bio refinery concept is critical to biomass commercialization.

6.0 NUCLEAR POWER: A RENAISSANCE?

Recently India came in the middle of a strong political debate on Indo – US Nuclear Agreement. The agreement for civilian sector emphasised on the potential development of nuclear power with assured supply of fuel from outside the country. Nuclear powers in India currently generated by 17 reactors remain within control of Government Agency; Indo – US deal is expected to open up the market for private investment. The current nuclear power generation of 4120 MW accounting for nearly 3% of India's electricity, is projected for a quantum jump. Even at current rate, The Nuclear Power Corporation Ltd., has made a profit of 37.2%; such lure of big profit has naturally attracted big corporate houses of "Reliance Energy" and "Tata Power to nuclear power prospects.

India has planned to set up four "nuclear power parks" to house 25 reactors capable of producing 45,000 MW of energy through a public private partnership. Estimated cost for a MW is pegged at Rs.6.0 crore, so an investment of Rs.2,70,000 crore would be required. The target is to reach 60,000 MW of nuclear power by 2026. Four sites have been selected viz. West Bengal, Orissa, Andhra Pradesh in east cost and Gujarat in the west. Table 9 below provides a scenario of possible development of nuclear power installation capacity in India till 2050.

Table 14.9:Possible Development of Nuclear Power Installed Capacity

Year	Unit	Scenario	
		Optimistic*	Pessimistic
2010	GWe	11	9
2020	GWe	29	21
2030	GWe	63	48
2040	GWe	131	104
2050	GWe	275	208

*It is assumed that India will be able to import 8000 MW of Light Water Reactors with fuel over the next 10 years.

Remarks: This estimate assume that the FBR technology is successfully demonstrated by the 500 MW FPBR currently under construction, India succeeds in assimilating the LWR technology through import, becomes completely self – reliant on PHWRs and develops the Advance Heavy Reactor for utilising thorium through an alternative route by 2020.

[Source: Department of atomic energy]

However, concerns of nuclear power as a safe source of renewable energy are being questioned. *(See "Nuclear Power Policy and Global Politics" in the book, chapter 12)*

Internationally, a nuclear winter was seen during last two decades. Although 440 nuclear reactors globally generate 16% of global need, the investment, safety during operation and disposal of radioactive waste plagues the scenario. Lake (see *e journal, European Perspective*) states that the 'US is on the verge of new nuclear power plants...this is the beginning of the third era, the renaissance of nuclear energy', but correctly laid down four principal challenges, which remain equally valid in India, USA or anywhere in the world.

- First, nuclear power remains economically competitive in the world energy market; in particular, energy companies must better control capital costs.
- Second, in order to satisfy the public's expectations of exceptional safety performance, current plants must continue to operate safely and future plants must continuously improve safety in expanding world markets.
- Third, nuclear power and its fuel cycle must be viewed by the public and by national leaders as sustainable; in particular, used nuclear fuel must be managed in a manner that is cost effective and safe for the extended period of time that used fuel remains highly radioactive, and the nuclear fuel supply must be extended for centuries in the face of depleting fossil fuels.
- Fourth, the nuclear materials from the fuel cycle must be protected from proliferation and misuse for non – peaceful purposes.

International efforts are now geared to develop next generation reactors – more economical, safer, sustainable and more resistant to proliferation of weapon grade plutonium; an Advanced Fuel Cycle Initiative to investigate advanced reprocessing and recycle strategies for used nuclear fuel is in the priority agenda. This is stated to be a part of "Global Nuclear Energy Partnership" (GNEP) in the coming years.

7.0 DISCUSSION

The Government of India promises that every village and every household in the country will have electric connection by the end of the next decade i.e. 2020. In reality, the past performances deny such hope. During last 10 years, about half of the promised increase in electricity generation has been achieved. Worst sufferers, due to lack of power, live in the hinterland. Officials now admit that it is an uphill task or impossible to add 90,000 MW of electricity by 2012.

Leave aside the cities, who suffer long hours of power cuts, the booming industrial growth will find it difficult to meet power demand. Simultaneously; 9% GDP growth, lead to the life style changes. Rural India's need so far was for non commercial power but rural India is now demanding commercial sources to reach a consumerist society (Jayanta Roy Chowdhury, "Solutions: Reality checks"; *The Telegraph*, 27 October 2007). Today India ranks 6th in the total energy consumption in the world but per capita consumption are only 20 per cent of global average. To meet the growing demand, renewables seems to be the only answer, keeping in view the global demand to cut carbon emission, India has to urgently plan and implement an alternate strategy for power generation. The new agreement, after the current Kyoto Protocol deadline is over in 2012, will invariably focus India and China for their future plan and commitment for carbon reduction.

Renewable Energy therefore has to play a major role in coming decades to meet the demand of

- Sustainability
- Pollution – free production
- Reduction in Carbon emission
- Meeting ever increasing need.

Water, Solar, Wind, Biomass, and Nuclear Power apart, use of hydrogen fuel is being researched. Hydrogen is the simplest, most abundant element in the universe but it does not occur naturally as a gas on earth. Enormous amount of investment is needed to develop appropriate technology and infrastructure to produce, store and distribute hydrogen for use in fuel cell vehicles and electricity generation: Best way to produce hydrogen from renewable sources, using such method is electrolysis to split water.

USA is playing a lead role in Hydrogen Fuel development. It has pledged $ 1.2 billion in 2003 for a Hydrogen fuel initiative. It is active in "International Partnership for the Hydrogen Economy" (IPHE) established in 2003 with 15 countries and European Commission. IPHE is focusing on developing Hydrogen codes and Standards to ensure safety of hydrogen power. (usinfo/states.gov/gi/acchive 2005/Apr/20 – 788575.html.)

In conclusion, one can visualise a better environment with development of all possible sources of renewables, keeping in view the demand of cost effectiveness, safety and health.

REFERENCES

Anon, 2006. Power options. *Dam, Rivers and People.* 3 (11 & 12): 21 – 27.

Anon. 2007. 'Solar Power Boom, *Dam, Rivers and People*, 5(10–11):1,8–11.

Business line, the Hindu, 2008. Push for MW – capacity solar power generation. January 07, 2008.

e journal USA. 2006. Economic Perspectives: Clean Energy Solutions. 11 (2):1–51.

Ghosh, A. K. and Das, S. K. 2003. Captive Energy Plantation for Renewable Energy in the Sundarbans, West Bengal, India. *International Energy Journal,* 4 (2):93 – 97.

Fairley, Peter 2002. Solar on the cheap. *Technology Review,* January – February 2002, 48 – 53.

Mondal, S.K. 2008. Renewable Energy. In Status of Environment in West Bengal –A Citizens Report. (ed. A K Ghosh) p – 215 – 220. ENDEV – Society for Environment and Development. Kolkata.

Pellerin, Cheryl. 2005a. Solar Cells Increasing use of Electricity from sunlight. usinfo/state.gov/gi/archive/2005/Apr 19 – 464962 html.

Pellerin, Cheryl. 2005b. Organic Material largest US source of Energy. http//: usinfo/state.gov/gi/archive/2005/Apr 21 – 101461 – html.

CHAPTER XV

Nuclear Power: Policy and Global Politics

1.0 INTRODUCTION

The World is facing a new energy challenge. The geopolitics of oil and natural gas and declining reserve of coal apart, the increasing global warming linked up with carbon emission are multiple factors driving policy makers to alternatives. Renewable energy sources – Solar, Wind, Hydel, Gasifier (biomass) are increasingly gaining grounds. The thrust on use of ethanol to produce biodiesel is yet another evolving issue. Clean energy is the new buzzword.

During this period of 21st century, the renaissance of nuclear energy is being hailed by the super power in USA. Nuclear energy had emerged in December 1942 when Enrico Fermi created a nuclear fusion reaction in an underground military laboratory in Chicago. Humanity's relationship with nature, changed for all the time. Fermi's discovery transformed warfare, ushered a nuclear age and opened yet another source of energy to produce electricity. Nuclear Power Plants produce clean energy without carbon emission, expensive but assuring and quickly spread over North America and Europe, on to Japan. No one talked much about the nuclear waste, which takes thousands or even millions of years for natural decay.

2.0 NUCLEAR WASTE : THE UNRESOLVED PROBLEM

It is said that from the beginning of the nuclear age, governments neglected wastes as best as they could (Lenssen, 1992). Old burial and storage site have proved leaky. Accumulation of irradiated fuels from Commercial Nuclear Plants increased from

12,601 Mt in 1985 in USA to 40,400 Mt in 2000. Canada, Russia, France, Germany, Sweden along with US, produced 40,191 Mt of irradiated fuel in 1985 by 1990 it increased to 84,000 Mt and by 2000, 193,000 Mt. (Lenssen, op.cit.). Radioactive waste contains unique blend of atomic structures called radioisotopes; each radioisotope has its own life span and potency to give off alpha and beta particles and gamma radiation, which can cause serious harm to living tissue leading to fatal cancer.

Most of the irradiated fuel in USA is stored in large pools of cooling towers located by the side of nuclear reactors but space limitations made many countries build large facilitics for waste disposal even away from Nuclear Power Plants. Some countries e.g., France and UK often reprocess the irradiated fuel but such a process can often leave behind radioisotopes created by splitting uranium atoms (cesium, strontium, technetium etc.), which can account for 97 per cent of the radioactivity of waste!

Several ideas of safe waste disposals have emerged but either abandoned or never seriously taken up. A recent report dated July 31, 2006 by the Committee on Radioactive Waste Management in U.K., shows the present status of nuclear waste disposal in the country. The report recommended a method in which solid waste-form could be kept in a container with an outer buffer of clay and crushed rock and placed in a host rock formation at a depth of 200-1000m underground; the report says that it will provide stable environment for radioactivity to decrease. By 2120, it is estimated that U.K., will have 478,000 cubic meters of high activity nuclear waste and no clear picture has emerged as what steps are being taken for the waste being generated since 1920. As a result, country wise protest came up when the U.K. Government announced its plan to build more nuclear energy power station (DTE, August 31, 2006).

In USA, nuclear power plants, built on the utopian idea of being so inexpensive that it was 'too cheap to meter'. The myth exploded with rising construction cost and safety performance, cumulating in the accident at the Three Mile Island Unit II plant near Middletown, Pennsylvania in 1979. US Nuclear Regulatory Commission (NRC) laid new rules for safe operations; this delayed the process of completion of plants 'during a time of double digit inflations'; several plants went bankrupt ending the first era of US

Nuclear Power. (Lake, 2006). But by 1990's, 103 nuclear power plants were producing 20 per cent of electricity in US at a much cheaper cost per kilowatt hour. By 21st Century, more than half of US nuclear power plants got extension of 20 years over their original 40 years licenses.

Table - 15.1: Technical Options for dealing with Irradiated Fuel

Method	Process	Problems	Status
Antarctica Ice Burial	Bury waste in ice cap	Prohibited by international law: low recovery potential, and concern over catastrophic failure	Abandoned
Geologic Burial	Bury waste in mined repository hundreds of meters deep	Difficulty predicting geology, groundwater flows, and human intrusion over long time period	Under active study by all nuclear countries as favored strategy
Long-term Storage	Store waste indefinitely in specially constructed buildings	Dependent on human institutions to monitor and control access to waste for long time period	Not actively being studied by gov'ts, though proposed by nongovernmental groups
Reprocessing	Chemically separate uranium and plutonium from fission products in irradiated fuel: decreases radioactivity by 3 percent	Increases volume of waste by 160-fold; poor economics; increases risk of nuclear weapons proliferation	Commercially under way in four countries; total of 16 countries have reprocessed or plan to reprocess irradiated fuel
Seabed Burial	Bury waste in deep ocean sediments	Possibly prohibited by international law: transport concerns; nonretrievable	Under active study by consortium of 10 countries
Space Disposal	Send waste into solar orbit beyond earth's gravity	Potential launch failure could contaminate whole planet; very expensive	Abandoned
Transmutation	Convert waste to shorter-lived isotopes through neutron bombardment	Technically uncertain whether waste stream would be reduced; very expensive	Under active study by United States, Japan, Soviet Union, and France

SOURCE: Worldwatch Institute, based on sources documented in endnote 32.

In absence of accidents, the issue of safe disposal of waste remains an highly unfinished agenda. But the worst Nuclear Power Plant disaster had taken place already in erstwhile Soviet Union at Chernobyl in 1986. Full account of impacts on air, water, soil and human life has never been published.

3.0 INDIAN SCENARIO

Nuclear Power Corporation of India Limited (NPCIL), from its plants at Tarapur, Maharastra, Kalpakkam, Tamil Nadu, Rawatbhati, Rajasthan, Narora, Uttar Pradesh, Kakrapar, Gujarat and Kaiga, Karnataka, is said to have generated 15560 million units of power during 2005-2006. NPCIL made a profit of Rs. 1713 crore during the same financial year. India's nuclear power plant programme started in 1969 and expanded to six states; capacities have been increased through expansion; further augmentation is foreseen in at least six of the sites viz., Maharastra, Rajasthan and Karnataka. A new power plant has been started in Andhra Pradesh (Kudankulup). Under Indo-US Nuclear Deal, more prospects are coming up.

In a paper 'Think again – US India Nuclear Deal' (*www.foreignpolicy.com/story/cms.php*) Graham, Tomero and Weiss (dt. July 24, 2006) serious aspersion have been cast over the possible diversion of nuclear fuel for civilian reactors for military purpose. The deal will allow the reactors to be off-limit for inspection. Nuclear fuel for power generation, the authors believe, will free India's uranium and plutonium stock for use in weapon. It doubts India's record of the past since India has secretly exploded bombs. Argument that 12.5 of power can be obtained from new nuclear power plant is also debated. India could be energy efficient and save 20,000 Mw of energy – the projected equivalent of country's nuclear power. US should remember India has not signed NPT.

However, major energy experts in India are already visualizing a different scenario. R. K. Pachauri, the Director General of TERI in a media article (*Times of India*, August 1, 2006) entitled 'Re-energizing India' expressed his deep satisfaction from a recent advertisement in *The Economist* seeking expression of interest in a 2400 Mwe nuclear power station in India. Pachauri stated that such an advertisement wasn't likely to have been placed by any organization five years ago. A major hurdle has been crossed, according to him, with the US Congress voting positively on the

US-India Nuclear Agreement. It offers 'a major change in opportunity that India can now avail of in the future development of energy supply'. Pachauri expressed his optimism that the nuclear power station would now be financed and possibly be established by the private sector; however no mention has been made about the mode of cost recovery from such heavy expenditure, or on the issue of public safety and dispcsal of nuclear waste.

In yet another report, Satish Chandra, Former Deputy National Security Advisor (*Deccan Chronicle*, 12 September 2006), questioned the statement of Prime Minister of India on the justification of nuclear deal in the Lok Sabha as an exercise in enlarging India's energy options. It has been pointed out that while the uranium resources can only yield 10,000 Mw of power for 30 years, but there was no mention of possible conjunctive use with India's thorium reserves which is largest in the world and which could yield 2,50,000 Mw of power. Chandra (op.cit.) has stated that it 'is clearly an instance of *suppressio veri, suggestio falsi*'. It has been correctly pointed out that while the risk of hydel power was focused by the Prime Minister but the far greater risk associated with nuclear power including those of nuclear waste disposal has never been mentioned. It is apprehended that once the Indo-US deal was concluded, Chandra argued, India's subservience to US will be enhanced many fold as continuance of benefit under it will be predictated on annual presidential determinations of India's good behaviour.

4.0 GLOBAL NUCLEAR ENERGY PARTNERSHIP (GNEP)

In February 2006, President George W. Bush of USA announced his plan to divide nations into two classes – nuclear fuel supplier and user nations. US President Asserted nuclear power as safe and clean, and could generate large amount of low cost electricity without emitting greenhouse gases. Mercifully, he mentioned 'we must dispose nuclear waste safely' but remained silent on the mode of disposal. He also insisted that nuclear technology and material must be kept out of reach of terrorist network and terrorist nations. To achieve his goal US President announced GNEP. US will collaborate with France, Russia and Japan from fiscal 2007. It professes to use innovative, advance reactors and new methods to use spent nuclear fuel. It has $ 250 million to launch the programme as a part of $ 632 million request. While US has not built a new reactor since 1970, till 1990 France has built 58 nuclear power plants during the same time supplying 78 per cent of energy. It aims at providing nuclear power fuel and construction of small scale reactors for developing countries. An international fuel supplying consortium is also envisaged.

Critics opined that Nuclear Power Plants would be too expensive to solve global warming. Nuclear Energy Institute (NEI) is however upbeat; it has spent millions of dollars in 1998-99 and succeeded in getting extension of licenses for the operating plants. NEI used global warming issues as a catalyst to garner public support for a nuclear renaissance. Counteracting NEI's claims, critics say that it is too expensive and unsafe, it produces toxic waste and encourage risk for nuclear proliferation. Nuclear power has survived, they argue, due to 'high government subsidies and to monopoly utilities that make the consumer pay its high cost'. It actually cost 10-12 cents per kilowatt hour as against 9 cents for solar thermal, 6-8 cents for oil, 5-8 cents for wind, 5-6 cents for coal, 2-8 cents for hydel; only solar photovoltaic is costlier at 15-20 cents than nuclear power consumers paid for nuclear energy extra two dollars per kilowatt hour! Original cost for nuclear power plants often exceeded by 600 %. Operational cost including constant retrofit, repair and maintenance has increased dramatically over the year. Dismantling any plant has become enormously costly.

The fate of nuclear power utility, after deregulation, therefore was most uncertain by end of 1990's; more plants were predicted to close down, no new plant was envisaged till 2020 and the share of nuclear power was predicted to decline to a mere 7 per cent according to a Department of Energy Report.

Cost of nuclear power in Europe and elsewhere have also increased. France that made biggest investment in nuclear energy had a debt burden of $ 30 billion on the national utility Electricile de France.

Nuclear power (1999) shared 5 per cent of global energy need, doubling that share would require 1320 new plants including hundreds in developing countries. To reduce greenhouse emission from coal fired plants, at least 2000 nuclear power plants would have to be built by next decades.

Nuclear Non Proliferation (NPT) professed control over the use of nuclear power for defense. But diverting civilian programme for military causes have seen in India, Pakistan, Iran, Israel and North Korea.

5.0 THE MYTH OF COMBATING GLOBAL WARMING

Life cycle analysis of nuclear power plant (from mining nuclear fuel material to nuclear waste generation) shows the process emit more greenhouse gases than coal based plants. Helen Caldicott, a well known expert opined that enrichment of uranium cause 93% per year of CFC gas, according to Department of Energy, USA.

The problems seem to be not affecting US or members of G 8 countries. Sunday Herald (July 9, 2006) wrote about an action plan for "global energy security" with objective to set up nuclear fuel plants in G 8 countries combined with the widespread sale of reactors to developing countries. The G 8 Plan professes it will

- Benefit all nation
- Augment robust nuclear non proliferation regime
- Install reliable safety and security system
- It will not allow developing country to enrich uranium fuel or to reprocess spent fuel to extract plutonium
- Developing countries will be permitted to run reactors but have to buy fuel, enrichment and reprocessing services from G 8.

It shows that G 8 is endorsing promotion of dangerous nuclear technology ever conceived – plutonium fast breeder reactor and reprocessing – while the same time condemning Iran and North Korea – says Green Peace International. Simply speaking it's a big-market, G 8 is targeting for.

6.0 NUCLEAR POWER AND GLOBAL WARMING

A MIT report estimated that by 2050 at least 1000-2000 new nuclear plants would required to reduce CO_2 emission.

It will lead to five times more nuclear waste generation, and require waste dump of the size of 'Yucca mountain' (of USA) every 3-4 years. CO_2 released worldwide due to electricity production is only 9 per cent.

Efficiency in energy use is one of answer to offset GHG due to electricity production. US spends $ 200 billion annually on electricity. Electricity saving innovations are available in the market but not widely used.

Use of renewable energy, most of which are cheaper than nuclear power, with massive investment and appropriate technology is the other option with exciting potential.

Australia, for instance, has enough potential of solar and wind power utilization, and liaison with America for nuclear power is termed dangerous.

7.0 CONCLUDING REMARKS

Nuclear Power Projects have been given a tremendous boost by USA and received support from G 8 partners. In the name of Global Nuclear Energy Partnership, President Bush of USA is making every effort to access a huge market in the developing world to sell reactors and nuclear fuel. This is ostenably to ensure energy security and lessen pressure on global oil and gas stock in the name of curbing global warming.

The discussions in the preceding parts reveal contradictions, anomalies and hyped-up features. The problems of initial investment, cost per unit of production, the Safety requirement, disposal of waste, decommissioning after life expiry remain unanswered and unresolved. This is in spite of the fact that latest report of the International Atomic Energy Agency has put lot of hope on the proliferation of the nuclear power plants in the coming decades (see Box).

Investment in renewable and enforcing energy efficiency, globally and in India could surely be an alternative strategy for energy security.

Box: International Atomic Energy Agency's Nuclear Technology Review, 2006: Excerpts

World wide there were 441 nuclear power plants (NPP) operating at the end of 2005, totaling 368 GW (e) of generating capacity, and supplying about 16 per cent of the world's electricity. This percentage has been roughly stable since 1986, indicating that nuclear power has grown at the same rate as total global electricity for 19 years.

Four new NPPs were connected to the grid in 2005 (two in Japan and one each in India and the Republic in Korea), and one laid-up plants was reconnected in Canada. This compares to five new grid connections (and one reconnection) in 2004 and two new grid connections (and two reconnections) in 2003. There were two NPP retirements in 2005, compared to five retirements in 2004 and six in 2003.

At of the end of 2005, eight plants had been completely decommissioned, with the sites released for unconditional use. Seventeen had been partially dismantled and safely enclosed, 31 were being dismantled prior to eventual site release, and 30 were undergoing minimum dismantling prior to long-term enclosure.

The net increase in global nuclear generating capacity during 2005 was 3259 MW(e).... there were three construction starts in 2005: Lingao – 3 in China (1000 MW(e)), Olkiluoto-3 in Finland (1600 MW(e)) and Chasnupp-2 in Pakistan (300 MW(e)). In addition, active construction resumed at two NPPs in Bulgaria. ... In 2004 there were two construction starts plus the resumption of active construction at two NPPs in the Russian Federation. There had been one construction start in 2003.

Current expansion, as well as near-term and long-tern growth prospects, are centred in Asia... of 27 reactors under construction worldwide at the end of 2005, 16 were in Asia. Twenty four of the last 34 reactors to have been connected to the grid were in Asia.

Japan has the largest nuclear power programme in Asia... Japan now has 55 reactors in operation and one under construction... Altogether Japan plants to add ten units to the grid by 2014, increasing nuclear power's chare of Japanese electricity to more than 40 per cent.

2005 has been a year of rising expectations for nuclear power. In March, high-level representatives of 74 governments, including 25 representatives at the ministerial level, gathered in Paris at a conference organized by the Agency to consider the future role of nuclear power.

The vast majority of participants affirmed that nuclear can make a major contribution to meeting energy needs and sustaining the world's development in the 21st century for a large number of both developed and developing countries. Among them were a number of countries currently without nuclear power programmes, such as Egypt, Indonesia, Morocco, Poland, Turkey and Vietnam. Among the challenges facing countries staring up a nuclear power programme is that of establishing the necessary supporting infrastructure including the legal and regulatory infrastructure.

Uranium prices, which had been low and stable for the previous decade and a half, continued their climb – from $ 25 / kg in 2002 to $ 112 / kg in May 2006. Uranium production has been well below consumption for about 15 years, and the current price increase reflects the growing perception that secondary sources, which have covered the difference, are becoming exhausted.

The world's current 441 operating nuclear power plants generate over 10,000 tonnes of heavy metal (tHM) of spent fuel each year. Less than one third is reprocessed for recycling as mixed fuel. The remainder is placed into interim storage facilities. About 1,90,000 tHM are currently in storage. Most are stored in water; but an increasing amount is in dry storage, which has become the preferred method for new away-from-reactor interim storage. Dry storage has the advantage of being modular, which spreads capital investments over time, and in the longer term, the simpler passive cooling systems used in dry storage reduce operation and maintenance requirements and costs.

However, early economic incentives for reprocessing and recycling diminished after the 1970s, partly because of the slowdown in nuclear capacity growth...because uranium resource estimates continually rose and partly because of secondary sources. Only one fast reactor, BN-600 in the Russian Federation, currently operates as a power reactor, and it uses not processed fuel, but fresh highly enriched uranium (HEU) fuel. India, however, began construction in 2004 of a 500 MW(e) prototype fast breeder reactor at Kalpakkam, and there is ongoing research in a number of countries.

[Source: IAEA; The Telegraph, Calcutta, August 2006]

Note: No where in the above text cost of establishing nuclear power plant, cost for operation and maintenance and the cost of nuclear waste disposal has been raised: Author.

REFERENCES

DTE, 2006. UK's nuclear mess: Waste doesn't make for welfare. *Down to Earth*, 62: 14, August\ 31, 2006.

Lake, James, A.,2006. The Renaissances of Nuclear Energy (http:// usinfo.state.gov/journals/ihes /706).

Lenssen, Nicholas, 1992. Confronting Nuclear Waste. p. 46-65. *In* State of the World. World watch Institute, Washington D.C.

National Power Corporation of India Limited. www.npcil.nic.in

CHAPTER XVI

Bio-Fuel

1.0 SEARCHING ALTERNATIVES

The current concern for global warming compelled the present system of governance to look for alternatives. The economy based on carbon-technology is expected to face significant crisis unless alternatives are offered. Undoubtedly, the politics of oil is playing a major role in such a scenario. The spiraling oil prices are one of the driving force, besides the happenings of climate change, to look for alternative fuel source. One such coming into the limelight in recent times is bio-fuel.

Bio-diesel is known to be very similar to the petroleum based fuel and can be used as a substitute. But it is regarded as a much cleaner fuel. It has no sulphur, no aromatics and about 10% built in oxygen that helps it burn fully. Even the poly aromatic hydro carbon which is known to be carcinogenic is found to be lower in bio diesel. Added advantage is that bio diesel needs very little or no engine modifications when blended with petroleum diesel up to 20% since most of the transport vehicles in India or elsewhere are run of diesel, the potential of bio diesel attracted global attention. In India, the initial support for bio diesel came from the Planning Commission way back in 2002-2003. The reason was not mainly to reduce carbon emission but to find affordable alternative fuel from indigenous sources. One has to remember that 70% of petroleum used in India is imported and the oil crisis has been plaguing the country with frequent jump in prices.

It is now well known that oil yielding plants such as Jatropha (*Jatropha curcas)* and Pongamia can be rich sources of such bio-fuel. Both the plants are known to be hardy and can even grow in poor soil condition.

Bio-diesel became an international issue in the era of increasing concern for climate change and global warming. Reducing carbon emission being the target, European Union recommended bio-diesel blending as a mandatory step. India, many believe, has competitive advantage. Technology is simple Oil is to be extracted from oil yielding plants (like Jatropha or Pongamia) and undergo a process of 'transesterification' converting it into fatty acid esters; the end product is used as a blend for petroleum derived diesel

Planning Commission of India has constituted an expert team who envisaged a pilot project to uses 400,000 hectors of land by 2006-07. (See Excerpts)

2.0 THE GLOBAL THRUST

In a recent media article, [US Department of State's 'e Journal' (July 2006)] it is stated that 'Bio-fuels, principally maize based ethanol, present the biggest investment opportunity in renewable energy in the United States for the next several years. Earlier such production was energy intensive but over the years the technology has changed for better efficiency and 'green house gas emission from such an ethanol' are 15-20 per cent lower than from gasoline. The demand has been growing rapidly and in 2006, nearly 18 billion litters of ethanol was projected. Capacity to produce ethanol is going to be increased by 50 per cent per year. Impact of bio-fuels can also be witnessed in auto manufacturing sector. General Motors (GM) a giant auto manufacturer has nine models than can run with mixture of 85 per cent ethanol and 15 per cent gasoline (Eckhart, 2006). Maize has been identified as a major source of bio-fuel production in USA.

Besides maize, conventional bio-diesel crops include soybeans, sunflower, rapeseed, palm oil and Jatropha. Soybeans appear to have gained ground as a commercial crop in more than 30 countries. The major soybean producing countries include Argentina, Brazil, China, Korea and USA. Soybeans also provide a major source of edible vegetable oil and protein for livestock feed use.

One has to remember that maize and soybean are also valuable agricultural crops contributing to global food and nutrition scenario. Controversy has already been seen regarding using land resources for growing crops for bio-fuel at the cost of food security. If carbon emissions during production of bio-fuel undermine 'positive gains' there is no point to advocate the same!

A recent report entitled 'Bio-fuel market Worldwide (2007-2010) provides latest updates. The *key finding* includes:

- Growing at the CAGR of more than 33% from the year 2007, world Bio-diesel production is likely to touch the mark of 12 billion liters by the end of 2010.
- Global ethanol market is expected to exceed 1, 20, 000 million liters mark by the end of the year 2020, with a CAGR of above 6.5% from 2007.
- Contributing more than half of the world bio-ethanol production, Brazil & US are the world's leading producer of the same. On the other hand, European Union is the leading Bio-diesel producer producing 92.68% of the global bio-diesel.
- Global production of Bio-fuel (ethanol and Bio-diesel) exceeded 48 billion liters in 2005.
- In 2004, ethanol made up for 44 % of the total fuel consumption in all (non-diesel) motor vehicles in Brazil, while in US, ethanol was being blended with 30 % of total gasoline sold.

In terms of geographical coverage, the report provides some information for every major continental area starting from Brazil, Canada, U S A in the Americas, Australia, China, Thailand, India in the Asia Pacific region, European Union, and other countries in South America, e.g., Argentina, or in Asia viz., Indonesia, Japan, Malaysia, Philippines; the information largely deal with Bio-diesel, Ethanol scenario and policy of each of this countries.

But the controversy persists questioning the very rationale of such global euphoria over Bio-fuel. Scientists agree that carbon emission offset by 'bio-fuel are lower than these achieved through afforestation' they even predict that 'bio-fuel plantations will end up emitting more carbon than they will help save'. *(*Rastogi; DTE 2007*)*. It is now estimated that conversion of sugarcane to ethanol only helps to serve less than 2 mg carbon per hector (1 mg = one tonne) as against 199 mg of carbon that is emitted during clearance of one hector of tropical forest. They have used a period of 30 years as a basis for comparing strategies because that's the period calculated to develop carbon free fuel technologies. After a final life cycle analysis of different crops they concluded that, 'forestation of an equivalent area of land would sequester two to nine times more carbon over a 30 year period than the emission avoided by the use of the bio-fuel'. (Righelato and Dominic, 2007).

The conflict further becomes critical when it seriously affects food security. 'Right to Food' being universally acknowledged, United Nations Special Rapportuer, Jean Ziegler, has called for a 5 years moratorium on bio-fuels, describing it as a "crime against humanity" (to convert food crop into fuel). Further it is argued that subsidies for growing bio-fuel crop should be reduced as has been done by the European Commission; the subsidy introduced only three years back in 2004, included a flat rate of • 45 (euros) per hectare up to 2 million hectare. While such subsidy may not seriously affect EU countries, spurt of activities in African countries can surely lead to such a crisis. Cereal prices have already gone up in Africa, compelling nations to import food. Investment in bio-fuel production by industrial houses has been extended to Forest Reserves in some African countries. In Uganda, such an attempt by Private Sector to grow sugarcane in 7000 ha., area in Mabira Reserve Forest, has been strongly opposed (DTE,2007)

But private investment and government support for bio-fuel has seen quantum jump across the World, as can be seen from the previously cited data.

3.0 INDIAN SCENARIO

The Bio-diesel Programme of the Government of India was started in 2003 at the initiative of 'Committee on Development of Bio-fuel' constituted by the Planning Commission. In August 2003, Ministry of Rural Development (MoRD) was given the task of acting as the Nodal agency. In January 2005, a Detailed Project Report (DPR) was prepared for pilot phase, involving plantation of *Jatropha* in 400,000 ha., of land. The Bio-diesel Purchase Policy of Government of India was announced in 2005. The National Mission on Bio-diesel was formed in 2006. All these activities show rather a fast pace. Simultaneously, at the state level, several states have initiated Bio-diesel programmes and announced state policies. Some of the more proactive states include Andhra Pradesh, Chattisgarh, Uttaranchal, and Rajasthan. Major advantages for India are projected in the potential use of degraded wasteland for growing hardy species like *Jatropha* and *Pongamia.* (Raju, 2006)

In India, ethanol production was initiated with the notification of Government of India on 'Ethanol Blending Programme' (EBP). This notification, effective from January 2003, made it mandatory to blend five per cent ethanol in petrol in nine States and four Union Territories.

Bio-diesel programme has often been projected as a package for rural employment and development in the country. It will also help reclaim wasteland – for which a mission was launched during the time of late Prime Minister Rajib Gandhi. So, one can argue that, besides assisting in the process of carbon reduction and cutting down crude oil import bill, bio-diesel programme can assist rural employment and wasteland reclamation. Chattisgarh state created a 'Bio-fuel Development Authority' in January 2005 with a target of covering 300,000 ha, with *Jatropha* plantation; by end 2007 and to increase it to 1,000,000 ha., by 2010. Nursery raised plants are to be distributed to the farmers initially free of cost –free up to 500 saplings; later, price can go up from 50 paisa to Rs.1.50 p. Private companies are encouraged to take on lease government land @ Rs.100 / ha. which can go up to Rs.1000 / ha. in the 8th year, (for 200,000 ha. of government land). *Jatropha* seeds from Chattisgarh are valued high as 'miracle seed'.

As a matter of fact, theft of germ plasm of *Jatropha* from Chattisgarh drew media attention in 2006. *Jatropha* germ plasm from Pendra and Sarguja was reported to have high yields and the material of research collection was allegedly taken away by ex faculty member of Indira Gandhi Agricultural University (IGAU), Raipur to Private Company DI Oil India Pvt Ltd. (Putul, February, 2006)

Seed apart, the availability of suitable land for *Jatropha* plantation in the country has been calculated (Ramakrishnaiah, 2006). Cultivatable wastelands and degraded notified forest area suitable for *Jatropha* cultivation in India has been estimated at 32.29 million hectares. Details of the categories are given in the Table 1 below. One has to remember that most of the degraded lands in the country belong to Common Property Resource (CPR) which are used variably for 'community pasteurs, 'community forests, panchayat land and common dumping and threshing areas,' besides for human habitation ('encroahment' by marginalized people!).

Table-16.1

Category of wasteland Suitable for Jatropha Cultivation	Area (in million ha)
A) Non-forest Culturable wastelands:	
i) Gullied/Ravinous-Shallow (mainly Community, Govt.)	1.03
ii) Land With Scrub (Government/Panchayts)	15.05
iii) Land With Scrub (mainly Community, Govt.)	3.74
iv) Saline/Alkaline-Slight (mainly private)	0.41
v) Shifting Cultivation-abandoned (community)	1.22
B) Degraded forest land: Degraded Forest – Scrub	10.84
Total	**32.29**

[Source: Ramakrishniah, 2006]

It is also suggested that a minimum area under the wasteland category, (suitable for plantation) has to be unidentified, such areas, under six different categories of wastelands, spread over states accross the country are tabulated below (Table 16.2)

Table-16.2

Category of Wasteland	Total Area (m ha)	Minimum area available in a state	States with minimum area under each category
Gullied / Ravinous - Shallow	1.03	>10,000 ha	(11 States) AP, Bihar, Gujarat, J&K, Jharkhand, MP, Maharashtra, Orissa, Rajasthan, Tamil Nadu, Uttar Pradesh
Land with Scrub	15.05	>1 lakh ha	(18 States) AP, Arunachal Pradesh, Assam, Chattisgarh, Gujarat, Haryana, HP, Jharkhand, Karnataka, MP, Maharashtra, Manipur, Meghalaya, Nagaland, Orissa, Rajasthan, Tamil Nadu, Uttar Pradesh

Land without Scrub	3.74	>1 lakh ha	(9 States) AP, Arunachal Pradesh, Chattisgarh, Gujarat, MP, Maharashtra, Meghalaya, Rajasthan, Tamil Nadu
Saline / Alkaline - Slight	0.41	>5000 ha	(5 States) Bihar, Maharashtra, Rajasthan, Tamil Nadu, Uttar Pradesh
Shifting Cultivation (abandoned)	1.22	>0.1 million ha	(3 States) Assam, Manipur, Mizoram
Degraded Forest - Scrub	10.84	0.2 million ha	(12 States) AP, Assam, Bihar, Chattisgarh, J&K, Jharkhand, Karnataka, MP, Maharashtra, Orissa, Rajasthan; Tamil Nadu
Total	**32.29**		

[Source: Ramakrishniah, 2006]

Ramakrishniah (op. cit) has also provided the projection of year wise demand for petroleum diesel in India vis-à-vis proposal for 5 per cent blending and the requirement of land area to make it feasible (Table 16.3). It is observed that only 12 mha of land can meet all such demand.

Table 16.3

Year	Diesel Demand (MmT)	Biodiesel Requirement (M.Ha.)	Jatropha Area (MmT)	Additional Area (M.Ha.) Biodiesel
2005-06	–	–	2.42	–
2006-07	52.33	2.62	2.54	0.12
2007-08	54.95	2.75	2.66	0.12
2008-09	57.69	2.89	2.79	0.13
2009-10	60.58	3.03	–	–
2010-11	63.61	3.18	–	–
2011-12	66.90	3.35	–	–

(Source: Ramakrishniah, 2006)

Further, the willingness of 14 state governments to pledge 1.72 mha for *Jatropha* cultivation, at the initial period, indicates the positive thrust into the bio-fuel scenario in India. (Table 16.4)

Table 16.4

(areas in ha)

S.No.	State	Forest Area	Panchayat Area	Total Area
1	Andhra Pradesh	15000	160000	175000
2	Assam	25000	–	25000
3	Bihar	50000	–	50000
4	Chattisgarh	50000	51400	101400
5	Gujarat	49800	–	49800
6	Haryana	50000	–	50000
7	Jharkhand	6600	65400	72000
8	Karnataka	–	500000	500000
9	Madhya Pradesh	–	53000	53000
10	Mizoram	14000	12000	26000
11	Rajasthan	60000	–	60000
12	Sikkim	5000	–	5000
13	Tamil Nadu	50000	–	50000
14	Uttaranchal	500000	–	500000
	Total	875400	841800	1717200

[Source: Ramakrishniah, 2006]

Jatropha plantation however, as claimed, is not too easy to undertake. The land quality and management practices are critical in commercial plantations. Intensive Management is essential for five to seven years but care may be needed to ensure commercial production even after three decades. It is now alleged that focus on

Jatropha has created a raw material shortage while other easily found plants are ignored (e.g. *Pongamia, Neem, Mahua* etc.). Need for high quality seed can be met by R & D investment – (DI Oil India of UK has reportedly invested Rs.80 crores towards such a project); other private sector companies are also eager to jump into the fray.

A media report states that 'The Government's *Jatropha* frenzy could cause serious social economic and environmental damage'. (Mishara and Awasthi, 2006). Degraded forest lands are often occupied by marginalized forest dwellers and their livelihood depends on these. Government of India's plan to put 11 million hectors under *Jatropha* actually means diversion of 7.0 per cent of the cultivatable land of the country including wastelands. Coarse cereals and other food crops can be grown in such land. Besides social cost, environmental cost may include irrigation water, application of fertilizer etc., when commercial cultivation is planned with an average seed yield of 5 tonnes/ha. Additionally, these plants may complete with other food crops demanding soil moisture.

A look at the "Ethanol" production from sugarcanes can now be taken. India, as one of the largest sugarcane producers of the World, has annual turnover of 300 million tons. Of this about, 60 per cent is used to produce sugar, 30 per cent to produce other sweeteners viz. *gur, khandsari* etc., rest being kept for seed production. Currently, as par industry sources, sugar production in India exceeds the demand of consumption and problem of carry forward stock is pegging the industry. Ethanol production is being looked upon as an alternate profitable usage of sugarcane. Blending ethanol with petrol in nine sugarcane growing states in India becoming mandatory can now help the cane producers. Doping petrol with ethanol is reportedly working successfully. But, industry feels the procurement price is unfair, Ethanol, an environmental friendly product, can be considered as oxygenate in place of methyl–tertiary–butyl–ether (MTBE). It is argued that if MTBE price is Rs.23.50 per liter with half the quantity of oxygen, the rational price for ethanol/liter should be Rs.47/liter compared to present supply rate of Rs.22.70 only. Agreeing that MTBE has other positive advantage, the price should at least be Rs.34/liter. Industry is demanding a steady procurement policy valid at least for 3-4 years, a Nominal Tariff Policy for Inter-State movement, Concessional excise duty, Surcharge etc. Expectedly, the government may relent to these requests in view of increasing oil import burden.

4.0 CRITICAL APPRAISAL

On the other hand, Bio-fuel frenzy is now being critically appraised at various levels. 'Bio-fuels implications for agricultural water use' has been an area of investigation for the researchers from 'International Water Management Institute' (IWMI) in Sri Lanka. A report from IWMI (October, 2007) states that India is likely to have severe water crisis and China will face shortage of level. It is predicted that India and China will account for 70 per cent of the projected global growth in fuel demand by 2030; with the current levels of efforts, bio-fuels may replace 7.5 per cent of fuel but at the cost of 30 million ha., of cropland and 180 hm of irrigation water. It may create a crisis in sustainable water use and affect food production. In the ultimate analysis, bio-fuel production may lead India and China to import food.

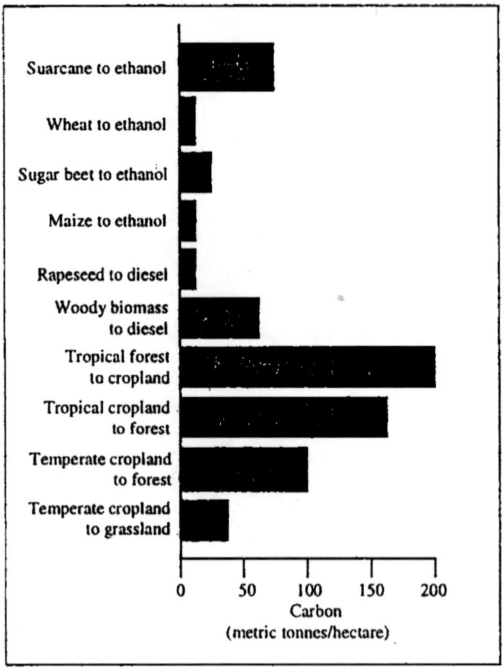

Fig. 16.1: Bio-fuels save little carbon

Source: Renton Righelatyo, Dominick V. Spracklen, 2007

Elsewhere in US, the issue of increased production of ethanol with the target of replacing 15 per cent of fuel by bio-fuel, by 2017 is being critically analyzed. A National Research Council (NRC) Report warned about the possible stress in Water quality and quantity to meet the demand of bio-fuel from Corn. Any planning in bio-fuel production, the report says, must be done with 'regional water assessment'.

Researchers from University of Leeds, UK, have shown that bio-fuels actually save little carbon. It is stated that 'Carbon, a natural forest can capture, is about nine times more than what bio-fuels can save during the same period' (Rastogi, 2007). The figure 16.1, below graphically represent the comparative carbon reduction/ absorption, in terms of metric tonnes per hector.

Coming back to India, it is felt that monoculture of *Jatropha* covering large tract of degraded forest land can never augment biodiversity. Although land use pattern for *Jatropha* stress on wasteland and cultivable fallow, shouldn't proper afforestation be expected to provide better habitat for biodiversity? (Fig 16.2)

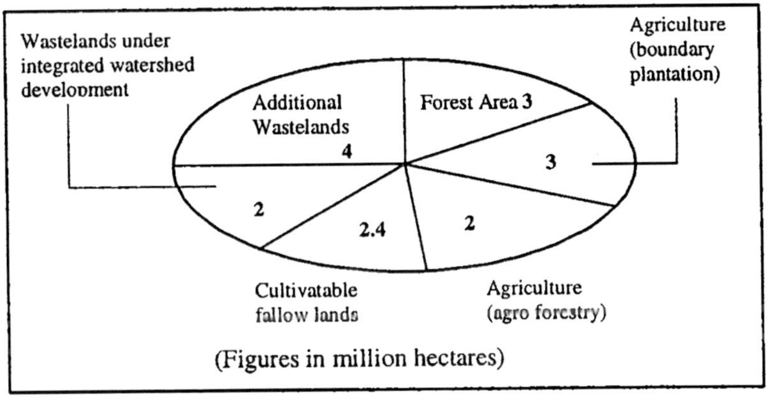

Fig. 16.2: Land-use pattern for jatropha

Source: www.jatrophabiodiesel.com

However, the bio-fuel boom is continuing even when questions about rationale of use of such fuel mix to curb carbon emission are being asked.

In Asia, Thailand is reported to have investigated the potential of Tapioca, Cassava and Palm Oil. Malaysia selected palm oil as major source. The Philippines looks to potential of Coconut oil and Indonesia to both Palm oil and *Jatropha.* Between 2003-2005 most of these Asian countries have announced their bio-fuel policy; suggested percentage of blending rates varying between 3 to 10. Even small countries like Bangladesh, private investors have reportedly pledged $4.5 million to set up 12000 liters/daily, ethanol manufacturing plant.

It is evident that Bio-fuel has been accepted as a future fuel option, however marginally, across geographical area. Policy support, development of technologies and market demand is accelerating the process. Critical issues like water demand, selection of marginalized level or wasteland, right agronomic practices, seed selection, yield potential etc. still remain largely unresolved. But the most urgent need is perhaps to resolve the issue of the appropriateness of converting land for bio-fuel plantation at the cost of food and water security. What the most "Developed Countries" can afford, can "Developing Countries" or "Least Developed Countries" afford to follow the same path?

Excerpts of the Recommendations of Report of the Committee on Development of Bio-fuels

In July 2002, the Planning Commission constituted a Committee on Development of Bio-fuels in order to accelerate the developmental pace in this emerging field of renewable fuels, and to study the various aspects in the area such as commercialization, prospects, quality projects, marketing strategies, research and development, etc. The committee has submitted its report to the Planning Commission. Here are the excerpts of the recommendations given by the Committee......

5.0 BACKGROUND

The envisaged Bio-fuels program is based on several compulsions and imperatives such as:

- the impending international oil crisis;
- steady spurt in oil demand for transport;
- domestic crude oil output reaching a plateau and the need to reduce dependence on imported crude;

- national commitment to provide clean air to the citizens as a part of their fundamental rights by enforcement of stricter emission norms;
- technical and economic feasibility of producing bio-fuels for blending with motor sprit and diesel on a large scale;
- implications of integrated bio-fuel development for upliftment of the poor, through reclamation of degraded land, drought proofing, provision of income and employment opportunities in plantation, primary processing and procurement/transport; and
- as a means to "green & rehabilitate" degraded lands on a sustainable basis.

These imperatives and the estimates of incremental demand of diesel and motor sprit in 2006-07 and 2011 suggest that a program to produce bio-fuels, and blend them with conventional fuels must be started right away.

From the current energy scenario, it is clear that India must move towards the use of bio-diesel for diesel. Such a move implies that the production of bio-diesel in 2011-12 and coverage of land with *Jatropha curcus* should be as follows:

- 3.35 MMT for blending @5% and coverage of area of 2.79 Mha
- 6.69 MMT for blending @ 10% and coverage of area of 5.58 Mha.
- 13.38 MMT for blending @ 20% and coverage of area of 11.19 Mha.

Blending of motor sprit with 10 per cent ethanol can be achieved if only additional capacity to dehydrate ethanol could be added. To achieve the above blending levels the following recommendations are made:

6.0 ETHANOL

- Blending of ethanol with motor sprit needs to make a very modest beginning.

- The target should be to raise the blending in stages to 10% by the end of the Tenth Plan (from the current level of five per cent blending in some states).
- If the industry finds it economically feasible, it should be encouraged to produce alcohol also from sugarcane juice directly in areas where sugarcane is in surplus.
- Restrictions on movement of molasses and putting up ethanol manufacturing plants may be removed.
- Imported ethanol should be subject to suitable duties so that domestically produced ethanol is not costlier than the imported one.
- Ethanol diesel blending requires emulsifiers and also poses certain storage and technical problems. The Indian Institute of Petroleum is working on the subject. Ethanol diesel blending should await the solutions to these problems.
- Buyback arrangement with oil companies for the uptake of anhydrous alcohol should be made.
- The reduced cost of production of ethanol, provision of incentives for new economic-sized distilleries incorporating state-of-the-art technology, and options for the integration of distilleries with sugar plants should be considered.
- Economics of ethanol production from other feedstock such as sugar beet, corn, potatoes, etc., should be studied. It may be left to the industry to use these raw materials for producing ethanol as and when it finds them economical.
- The micro-mission on R&D proposed to be set up under the National Mission on Bio-diesel will look after the areas relevant to the ethanol sector also.

7.0 BIO-DIESEL

- A National Mission on bio-diesel should be launched immediately with the object of producing bio-diesel required for blending to the extent of 20% with HSD by 2011-12.

- In phase-I, a demonstration project may be taken up on 400,000 hectares in eight states.
- On the overall basis, the employment under the project will directly enable 5.50 lakh rural poor families to rise above the poverty level.
- The National Mission on Bio-diesel may be based on *Jatropha* as it has many advantages over other species including *Karanja (Pongamia Pinnatta).*
- The demonstration project may be completed by 2007. The next phase of the National Mission should be people driven and should involve a self sustaining expansion of plantation and setting up of corresponding facilities for seed collection, oil extraction, esterification, etc. The Government should act mainly as a facilitator providing incentives as may be necessary.
- Efforts should be made to get external funding for Phase II of the National Mission. The Government may fund the demonstration project under the National Mission.
- As the implementation of the demonstration project makes progress and bio-diesel starts becoming available, a beginning should be made with 5% blending in areas where the production of bio-diesel is taken up by the year 2005.

8.0 MICRO-MISSIONS UNDER THE NATIONAL MISSION

The National Mission may be implemented in a mission mode. The Demonstration Project may consist of a number of micro-missions. These include micro-missions on plantation on forest and adjoining lands and on non-forest lands – implementation by NOVOD Board; a micro-mission on plantation not covered by the earlier two micro-missions (implementation by Ministry of Rural Development); micro-mission on procurement of seed and oil extraction; micro-mission on trans-esterification, blending and trade; and micro-mission on research and development.

9.0 FINANCING

For the demonstration project, the total cost has been estimated to be Rs. 1,496 crore. For the plantation component (Rs. 1,200 crore), the funds need to come from the Government. For the component of setting up a Seed Collection Centre and Oil extraction unit, the funds could be a mix of entrepreneurs' own contribution (margin money), subsidy and loan from NABRAD and SIDBI in the ratio of 10:30:60. The amount of subsidy will again have to be provided by the Government.

In the last year of the demonstration project (2006-07), on the basis of experience gained, a project will be formulated for Phase II of the National Mission. The villagers are expected to grow *Jatropha curcus* plantation on their fields as an agro-forestry crop. A scheme of margin money, subsidy and loan may need to be instituted. Companies with land could be encouraged to undertake *Jatropha curcus* plantation. Funds for plantation in degraded forests through JFM could come from the JMF members.

It is also noted that since the bio-diesel program will address global environmental concerns, it is likely to attract the support of bilateral/multilateral agencies.

10.0 INSTITUTIONAL ARRANGEMENTS FOR THE NATIONAL MISSION & DEMONSTRATION PROJECT

A co-ordination committee under the Chairmanship of the Deputy Chairman of the Planning Commission and Steering Committee of officials to be served by a compact cell in the Planning Commission may be set up. In the beginning there is a need for flexibility as a very rigid legal regime may hamper the development of bio-fuels in India. Hence it is proposed that a separate legislation on bio-fuels need not be considered at this stage and the necessary legal requirements may be met by using the already available status.

REFERENCE

DTE, 2007. Bio-fuel Plans Clipped. *Down to Earth.* P-11, November 2007.

Eckhart, Micheal.2006. Renewables: Looking Towards Inexhaustible Energy e Journal USA, *Economic Perspectives.* 11 (2):18-23.

Mishra, Saurav and Avasthi, Kirtiman, 2006. Jatropha Jam. *Down to Earth.* September 15, 2006, New Delhi.

Putul, Alok Prakash. 2006. Biopiracy; Jatropha germplasm theft in Chattisgarh. *Down to Earth.* February 28, 2006, New Delhi.

Raju, K.V. 2006. Bio-fuels in South Asia: An Overview. *Biofuels India.* 4 (3): 2-5.

Ramakrishnaiah, D. 2006. Land Availability for Biofuels Plantation. *Biofuels India.* 4 (3): 6-9.

Rastogi, Archi; Biofuel Hunger, *Down to Earth.* November 15, 2007: 28-29.

Righealto, Renton and Spareklen, Dominik,V., Environment: Carbon Mitigation by Biofuels or by Saving and Restoring Forests? *Science.* August 17, 2007: 902.

CHAPTER XVII

LOOKING FOR ALTERNATIVES: BIOPESTICIDES

1.0 INTRODUCTION

The era of chemical revolution ushered the miracle of chemical control of pests and diseases in Agriculture. Food security being a vital issue for burgeoning millions specially in the developing world, use of chemicals in agriculture became a phenomenon. The increasing impacts of these chemicals in nature, natural resources and public health caused serious concern with the now famous publication of Racheal Carson's - 'The Silent Spring' in early 1960's. During last five decades serious thoughts have driven researchers to a slowly developing area of science, that of Biological Control. Biological control methods, however, were known to the farmers at the grassroot level. Such methods were innovated out of traditional wisdom and inherited knowledge and practices but never formalized and methodically tested on a scientific protocols. But post 1960's the proven fear of chemical contamination perhaps led to more concerted efforts to develop and use Biopesticides. Nagarajan (2004) while defining Biopesticides included microbes, transgenics and products derived from plants and animals. There may be serious debate whether "transgenics" should be included as it may produce controversial impacts in biodiversity rich areas. Classically products of plant and animal origin can safely be included to avoid such controversy.

One of the limitations of Biopesticides is its character of being non target specific. Biopesticides also may behave differently due to differences in edaphic conditions. The impact of use of Biopesticides may not be immediate and dramatic and results may be slow to emerge. Botanical biopesticides have limitation of shelf life, high specificity, lack of easy access and differential qualities. (Nagarajan, 2004). The botanicals (which come from plants of Asteraceae, Labiatae, Malvaceae, Meliaceae, Rutaceae, Compositae, Solaceae etc.), but they have a long history associated with traditional knowledge. Modern science has fine tuned and validated many of the properties of botanical pesticides.

2.0 BIOCONTROL

However, other biocontrol agents, bacteria, viruses, fungi, nematodes and parasitoids and predators also play a significant role in the scenario; but obviously they can not be categorized as Biopesticides except perhaps the products derived from bacteria, viruses and fungi. EPA recognized at least 195 formulations / derivatives or organism belonging to the category of Biopesticides largely of origin from the above mentioned groups (www.epa.gov/pesticides/biopesticides/productslistsindex.htm); of these strains of *Bacillus thuringiensis* occupy 16 positions but also include 17 other sub species under serial no. 17-84. The list covers a diverse range of plant material from onion to kaolin, mint oil, castor oil, cedar wood oil, oil of citronella, oil of eucalyptus, oil of orange, oil of lemon grass, red pepper and of coarse neem. But the biocontrol agents also include other bacteria (*Pseudomonas isolate)*, Agro bacterium strains and a whole range of chemicals like CO_2, sulphur and list of other derivatives.

It is reported that more than 4.00,000 chemicals have plant based origin, of which 10,000 are secondary metabolites with potential to be used as insecticides. The chemicals from plants include terpenoids, alkaloids, glycosides, phenol, polyacetylenes and others.

The present statistics revealed that at least 2400 plants species have proven pesticidal potential; of these 1005 have insecticidal, 384 antifeedent, 297 repellent, 27 attractant and 37 growth inhibiting properties. How many of these have been researched to the stage of commercial production of biopesticides?

3.0 BIOPESTICIDE INDUSTRY AND MARKET

The expansion of biopesticide industry however still remains limited. It is also noteworthy that most of such units in India are located in southern part. An estimated 410 units were operating in India till 2004 (Singhal, 2004). However private investment accounts for only less than 30 (130 units) per cent, while others are in Public sector. But even with 30 per cent units, the corporate efficiency in private ownership run units is evident in terms of supply even to remote areas and also in terms of productivity. It is alleged that biopesticide industry in India remains unconcerned about quality control.

On the consumption side, the profile remains extremely poor. It has reportedly increased from 2001 by 2.5 per cent in 2003 with an ambitious target of reaching 12-15 per cent within next two years i.e., 2005. The results obtained are not yet available. The problem of consumption deficiency is not related to manufacturing and supply but perhaps linked with poor marketing, proven results, easy availability. It is reported that insufficient demand led to under capacity utilization of existing production units. Singhal (op.cit) recommended a series of tariff and non-tariff measures to boost the industry. Important recommendation includes allocating 50 per cent of state plant protection budget for biopesticide, waiving customs and excise duty and harmonizing of sales tax at the state level.

Of the non-tariff measures, seed treatment by biopesticides to make it safe before planting, setting up Good Laboratory Practices (GLP) for biopesticide industry, imparting training on use of biopesticides and providing incentive for the farmers using biopesticides as a part of organic food production, are some of the major recommendations.

The other biological control agents, viz., insects parasitoids, insect predators, mite predators and pathogens can act as effective alternatives to chemicals. Currently 10 species of egg parasitoids mostly *Trichogamma*, two species of egg larval parasitoids, 18 species of larval parasitoids besides group of entomophilic nematodes, 6 species of nymphal parasitoids and two species of pupal parasitoids are available for standardized mass production; additionally, two species of predatory mites and 10 pathogenic organisms or strains are available for mass production. Such a process needs to be extended through extensive research, to other identified biocontrol agents.

4.0 AGRICULTURAL SECURITY AND CROP DAMAGE

In a recent estimate, India has been projected as a country contributing more than 20 per cent of paddy, 11.3 per cent wheat, 27.5 per cent groundnut, 22.5 per cent sugarcane, 53.3 per cent jute, 14.6 per cent cotton and 7.8 per cent tobacco to the International Agricultural Crop production scenario. The food grain production grew four fold from 52 million tonnes in 1951-52 to 195 MT by the end of 20th Century. But with all these impressive figures, the loss due to damage caused by pests, diseases, weed invasion has been estimated at 20-30 per cent with an estimated loss of 200 billion Indian Rupees. The traditional method of chemical control obviously could not solve this problem of staggering loss of national revenue. The problem of chemical contamination of crop and crop products haunt exporters in view of an environmentally conscious importing country. Within India, the Government of India appointed expert committee to probe and suggest measures against pesticidal contamination of environment. But chemical contamination still remains a major critical area. Some pesticides have been banned in recent time (like DDT) in agriculture but still is officially used in Public Health.

On the other hand, organic food market has increased significantly outside India. An estimate in late 1990's puts it at a value of 27 billion US dollars. Agriculture and Processed Food Development Agency (APFDA) had a target of exporting organic vegetables, fruits - raw and processed to the tune of Rs. 55,000.00 million annually during 11th Five Year Plan. The back up support however was seriously lacking in terms of production of biopesticides. The experience of Indian Tea exporters in European market compelled some of the high quality tea producers to switch over to chemical free tea gardens. The Department of Biotechnology (DBT) claimed to have made concerted efforts by supporting 200 R & D projects in the area. The projects were funded to develop better biopesticides , commercially viable mass production of biocontrol agents and to conduct frontline field trials in different agroclimatic zones. But apparently results did not finally deliver the anticipated products.

The constraints and limitation of effectively implementing Integrated Pest Management has been deliberated over years. In final analysis, biopesticides have not attracted the attention of policy makers at the Agricultural sector to the extent that a time targeted programme is launched. Such a policy has to be implemented through Private – Public participation but in a Technology - Mission mode. The cost benefit ratio in biopesticide industry seems to be less lucrative to the pesticide lobby. But in the national interest, efforts to provide technology support, field training, organizing outreach programme for farmers and initiating a method of providing incentives and also disincentive are urgently needed.

A chemical free world in the food sector still remains a distant dream. But potential of reducing chemical contamination of food and food product has increasingly become visible. Farmers in the fields in Bengal cultivate own eatable crop products without chemical, being aware of ill effects on health but are forced to use of chemicals in the products meant for sale. Recently, Government of India has proposed to table the "Food Safety and Standard Bill" replacing the present "Prevention of Food Adulteration Act'. But will the Bill go to the root of safe food requirements? The political will seems to be sadly lacking.

Box – Pesticide Poisoning

The Pesticide contamination in India is regulated by the Prevention of Food Adulteration Act (PFA), 1954 by the Ministry of Health. The registration of pesticides is controlled by the Central Insecticide Board (CIB) of the Ministry of Agriculture, through the Insecticide Act, 1968. The linkage between these two ministries may be traced to the fact that the post of the Chairman of the Registration Committee is filled up by the Director General of Health Services.

But all the regulatory control seriously lack the appropriate standard to be set up for 'Average Daily Intake' (ADI) permissible for Adults and Juveniles. Maximum Residue Level (MRL) on food commodities are also never notified by the CIB. These two serious lapses can lead to a critical situation. Elsewhere the registering authority sets up ADI and MRL to safeguard the public health. Interestingly, the "Prevention of Food Adulteration Act" mandates MRL but till 2003 only for 71 out of 180 registered pesticides, MRL was notified. MRLs are fixed up by a 'Pesticides Residue Sub Committee' of the Central Committee for Food Standards without any references to ADI !!

The story does not end here. The Indian Agricultural Research Institute (IARI) is supposed to monitor Pesticide Residue through an "All India Coordinated Research Project" (AICRP). But it is not empowered to enforce standards. And in many cases like branded baby food and milk, standards are non-existent! Data generated by the project is normally kept secret.

In spite of all the limitations and constraints, the available data on pesticide contamination is revealing. A decadal study (1986-1996) found 55 per cent samples of Fruits and Vegetables contaminated with pesticides, the dominant ones being Monochrotophos, Methyl parathion and DDVP, with 26-31 per cent of samples above MRL (AICRPPR, 1999).

Another ICMR led study showed 'majority of milk samples are contaminated with residues of either DDT or HCH or both, and invariably .... exceed their prescribed MRL level'.

Its not only fruits and vegetables or milk and milk products, the entire range of cereals and grains, pulses, spices, honey showed variable level of contamination averaging an alarming 54.4 per cent of samples tested between 1965-1988 (DTE, December 31, 2003).

A report from Pollution Monitoring Laboratory of CSE (March 2005) further showed much higher Lindane residue (605 x), BHCH (129 x), Heptachlor (15 x), DDE (35 x), DDT (188 x) in the blood samples of Punjab farmers when compared with US Study Results.

[The data used for the box is largely based on "A Refreshing Guide to Food Safety", *Down to Earth* issue of December 31, 2003 and "Bhola Singh is Dead", *Down to Earth*, issue of June 15, 2005, and other sources.]

REFERENCES

Anon, 2000. The Hindus Survey of Indian Agriculture, Chennai, The Hindu Group.

Anon, 2006. your Goose is cooked: And the food safety bill has all the ingredients to do it. p. 39-42. *Down to Earth*, June 30, 2006.

Nagarajan, S., 2004. Role and limitations of biopesticides in sustainable agriculture. p. 1-11. *In* 'Biopesticides for Sustainable Agriculture : Prospects and Constraints'. (ed. Nutan Kaushik). The Energy and Research Institute, New Delhi.

Singhal, V., 2004. Biopesticides in India. p. 31-40. *In* 'Biopesticides for Sustainable Agriculture: Prospects and Constraints'. (ed. Nutan Kaushik). The Energy and Research Institute, New Delhi.

Upadhyay, R. K., Mukerji, K. G., Chamola, B. P., Dubey, O. P., 1998. Integrated Pest and Disease Management. APH, New Delhi.

CHAPTER XVIII

Greening of Indian Cities

1.0 INTRODUCTION

The urban planning process for long remained a domain of architects, town planners and engineers. The urban habitats of India remained 'chance directed and chance erected'. The concept of improved quality of life was confined to solid waste disposal, wastewater treatment, water supply, roads and transport, access to institutional, commercial and trading facilities. The need for integrating environmental management by way of roadside plantations, setting up of parks, gardens, open space and maintenance of urban wetlands didn't get adequate focus even in the first colonial capital – the City of Kolkata till the advent of Calcutta Improvement Trust (CIT) and that too after the capital was shifted to New Delhi in 1911-12. Green City is not merely for a visual, aesthetic pleasure but to give the citizens a chance to breath cleaner air, desist serious cardio – pulmonary diseases and help the citizens to appreciate nature and biodiversity. This was sadly lacking.

In the past, towns were planned along the riverine course to provide access to riverine transport and avenues for trade and commerce. The small township around fort area in many parts of India grew up to serve the royalty and lacked any planning process. The advent of colonial rule witnessed development of many cantonment areas for military services often with large expanse of land and greenery. However, except for a limited few cities remained outside the orbit of planned bio-aesthetic process.

2.0 THE GENESIS OF URBANIZATION

There is no chronicled record of green city concept in Indian sub continent. In the post independence period, towns and cities witnessed an unprecedented growth in some states and the process of urbanization was accelerated with every five-year plan since 1950.

The trend of urbanization between 1901-1951 was 17.29% in India but the process was faster in West Bengal at 23.89%. Surprisingly while Indian urbanization reached 25.71% by 1991, the rate of growth of urban areas in the state stood at 27.48%. But even then, an effort for integrating bio-aesthetic into the planning process was hardly visible in India or in the state. [One has to remember that urban centers are classified into six class categories viz., I : 1,00,000 and above population, II: 50,000 to 99,999, III: 20,000 to 49,999, IV: 10,000 to 19,999, V: 5,000 to 9,999 and VI: below 5,000]

The greening of urban centers, however, cannot be described as a formidable task when one considers that less that 5% of the total geographical area in India has been officially urbanized. All that perhaps is needed is a process of appropriate planning to integrate with the land use development and control plan within the broader framework of urban plan. It may be worthwhile to note that while only one city was having a population of more than one million in 1901, the number of such cities have increased to 23 by the decade of 1990's with consequent changes in the environmental quality of urban centers.

3.0 URBAN GREENING

Calcutta was the first capital of colonial India. A. P. Benthall (1946) published an excellent account of the trees of Calcutta and its neighbourhood. It enlists 276 species of which 69 were considered 'native' or naturalized in the neighbourhood besides 91 other indigenous Indian Plant, 50 species from other parts of Asia, 14 native to Africa, 42 of America, 9 from Australia and one from South Sea Islands. The greening of Calcutta had at least a systematic approach during colonial period especially in the central district area of BBD Bagh in the north to Park Street in the south; this was predominantly the habitat of colonial rulers. From bio-aesthetic point of view, the flowering pattern and their colour blooms were considered but obviously there was no planning for selection of species to act as absorbent of air pollutants (Ghosh, 1988) as the problem was unknown at that time. The process of creating new parks and gardens along the newly developed roads by CIT can perhaps be taken as a first serious step to make the city green, along with the venture for new road side plantation programme. This process continued, interruptedly till 1960's.

Elsewhere, urban greening was initiated in the new township of Chandigarh and in the city of New Delhi in the post independence period. The planned greening programme in New Delhi has gifted the city with highest percentage of parks, garden and open space amongst four major metropolises of India. Randhawa (1978) may be credited for propagating such a concept; Randhawa (op.cit) used the term bio-aesthetic planning following Professor Lancelol Hogben. The Green City concept however was first used in U.K. in 1950's. But it never had influence on urban planning process in India to a significant extent.

3.1 Planning Green City

Today, urban greening, as stated can not only serve the purpose of added aesthetic and visual pleasure but can help to reduce the impact of air pollution on human health. A brief indicator of dust collecting ability of tree species (depending on plantation in appropriate location) is given below. (Table 18.1).

Table – 18.1 : Dust Collecting Capacity

Sl. No.	Common Name	Dust Collection (gm/m^2 of leaf surface)
1.	Teak	5.35
2.	Sal	4.50
3.	Arjun	4.49
4.	Mango	4.05
5.	Kanchan	3.90
6.	Palas	3.05
7.	Neem	2.92
8.	Imli	2.08

The tree species therefore act as sinks for air pollutants on account of large surface area of leaves absorbing pollutants. Tolerance limits of tree species however may differ e.g., *Neem* is tolerant to sulphur di-oxide while *Sisso* is sensitive. As such trees may also be used as bio-indicators for air pollution (Table 18.2).

Table – 18.2: Air Pollution Tolerance Index

Sl. No.	Common Name	Air Pollution Tolerance Indices
1.	Siris	62
2.	Neem	38
3.	Amla	38
4.	Imli	30
5.	Mahua	18
6.	Bel	18
7.	Mango	11
8.	Jam	10
9.	Palas	10
10.	Teak	10
11.	Babla	9
12.	Chatim	9
13.	Jhau	7
14.	Sisso	6

The urban greening programme therefore has to combine the need for meeting the demand of reducing pollution load and increasing aesthetic values. A project report was prepared for greening of Calcutta Metropolitan Area, on the tercentenary year in 1990 (Dept. of Forests, Govt. of West Bengal, March 1990). The plan envisaged site specific plantation under five broad categories with desirable characteristics and recommended plant species viz., (i) Parks, Gardens, Open Space and Housing Complex, (ii) Industrial area, (iii) water bodies – Embankments, Canal side and River side and (iv) the Roads (both wide and narrow) and (v) Slum areas (Table 18.3).

Table – 18.3: Site specific Plant Recommendations

Sl. No.	Site	Desirable Species and Characteristics	Recommended Plants
1.	Parks and Gardens, Open spaces and Housing Complexes	Evergreen ornamental tall trees with spreading crown along the periphery, planting medium height trees in groves, ornamental climbers and creepers on topiaries, provision for small lawn for green ambience	Tree : *Samanea saman, Pterospermum acerifolium, Swietenia* spp., *Delonix regia, Jacarandra ovalifolia, Millingtonia hortensis, Ficus* spp., *Schiliechera oleosa* Small trees & shrubs for grove: *Callistemon citrinus, Caesalpinia pulcherrima, Muraya paniculata,* Ornamental bamboos, *Cestrum nocturnum* Climbers & creepers: *Bignonia venusta, Vernonia* spp., *Bougainvillea* spp., *Alamanda purpurea, Petrea volubilis*
2.	Industrial Areas	Closely planted strip plantation with evergreen tall and large crowned trees with thick coriaceous leaves.	*Polyalthia longifolia (pendula), Eucalyptus* spp., *Poplars, Azadirachta indica, Alstonia scholaris, Acacia mangium, Acacia aurculiformis, Casuarina equistifolia, Spathodea campanulata, Nerium indicum, Pongamia pinnata, Thespesia populnea*
3.	Water bodies, Canal and River banks	Evergreen, water tolerant trees mixed with fruit species to attract avifauna	*Syzygium cumini, Terminalia arjuna, Lagerstromia flos-reginea, Thespesia populnea, Ficus* spp., *Trema orientalis, salix* spp.
4.	Roads		
a.	Wide to moderately wide roads with overhead interference	Evergreen and ornamental trees- heght not exceeding 3-4 m.	*Bauhinia* spp., *lagerstroemia thorelli, Cassia bicapsularis, Cassia fistula, Caesalpinia pulcherrima, Cordia sebastina, Nyctanthes arbortristis*

b.	Wide to moderately wide roads without overhead interference	Evergreen and ornamental tall trees with spreading crown	*Putranjiva roxburghii, Delonix regia, Peltophorum terocarpus, Samanea saman, Swietenia* spp., *Ficus* spp., *Azadirachta indica, Michelia champaca, Alstonia scholaris, Jacarandra ovalifolia, Polyalthia longifolia, Grevellia robusta, Tabebuia crysantha, Tabebuia rosea*
c.	Narrow roads with moderately available space for planting	Evergreen ornamental, medium height trees and foliage shrubs	*Murraya paniculata, Ixora* spp., *Caesalpinia pulcherrima, Tecoma stans, Cassia bicapsularis, Lagerstroemia thorelli, Dombeya wallichii, Calliandra haematocephala, Tabernaemontana divaricata, Aganosma caryophyllata*
d.	Narrow roads with no space for planting	Ornamental climbers and shrubs	*Bougainvillea* spp., *Bignonia vinusta, Alamanda antigonon, Petrea volubilis, Holmskoildia sanguinea, Quisqualis indica, Clerodendron inerme etc.*
e.	Central Boulevards	Ornamental shrubs and foliage	*Ixora* spp., *Tabernaemontana, Nerium indicum, Lantana camara, Lagerstromia lancasteri, Jasminum multiflorum, Hibiscus, Jatropha, Euphorbia, Thevetia peruviana, Crotons, Acalypha, Aralia, Pedilanthes* etc.
5.	Slum Areas	Small sized flowering trees and shrubs where space is available	*Cassia bicapsularis, Bauhinia* spp., *Tabebuia* spp., *Saraca indica, Cassia fistula, Ixora, Cestrum nocturnum, Dombeya, Thevetia peruviana, Nerium indicum, Taberneamontana divaricata, Gardenia jasminoides, Nycthanthes arbortristis, Hibiscus mutabilis, Ichnocarpus fruitescence*

The 'Operation Green', Calcutta started in 1990-91 but ended abruptly. This is indicative of lack of commitment on the part of governance. With a spurt of activities in 21st Century, the city has suffered significantly from loss of green cover. The efforts to compensate and or to transplant remained half-hearted. In absence of an appropriate law for conservation of trees in non-forest areas, the loss of urban green has attracted severe criticism both from members of the public and press. A recent proclamation to restore the green at a cost of more than 27 million rupees is yet to be implemented. This is claimed to be the first coordinated efforts of 5 major stakeholder departments; it is aimed at planting 20,000 avenue trees, creation of 25 acres of urban forests, 20,000 seedlings, planting shrubs on boulevards etc. Forest Department is to supply 2,75,000 saplings over next three years. The Bio-aesthetic plan for Calcutta may finally come up by 2010.

While the green city concept is not confined to the roadside plantation, the assessment of green cover in the demarcated areas of Parks, Gardens and open spaces is also essential. The sheer number of Parks may not be very convincing in view of degraded green cover in many of the designated areas. A recent assessment of urban green in 141 wards of Calcutta reveals existence of 290 parks and gardens but 34 wards out of 141 do not have a single park; and less than 9% of the greenery have good canopy cover and nearly 55% have hardly any green cover; the existing green cover is composed of 170 species of flora of which 98 species are indigenous and the rest are exotic (Report on Parks and Gardens of KMC, ENDEV, 2002). Only 5.5% of the city area is occupied by Parks, Gardens and open space as against recommended 15% for a city of the size of Calcutta.

Obviously, an analysis of Green City concept can not be confined to Calcutta. Of the three other metropolises, New Delhi has more parks and open space and green cover than any other. The onslaught of South Delhi ridge has been stopped through intervention of the civil society. Advantage of Mumbai and Chennai on the other hand lie in the fact that Borivili Sanctuary and Guidney National Park come within the urban limits of Mumbai and Chennai respectively. Mumbai had 0.5 million trees in 1998 but increased its green cover to 0.8 million trees in 2005. With 12 million people, ideal coverage should consist 4.0 million trees. The Greater Mumbai Tree

Authority stipulates planting of two trees for every tree felled. Pune had 1.2 million trees in 1991, which has now increased to a cover of 3.4 million, which indicates a positive trend towards a green city concept.

Besides four major metropolises, the Garden City of Mysore and Green City of Bangalore were centres of attention over the years. Unfortunately the process of intensive urbanization under open market economy, cities like Bangalore have lost earlier green cover and consequent change is local salubrious climate.

Integrating advantage of having natural green cover through intelligent planning and augmenting the green cover through planned plantation may be the most desirable goal. While greening of the city may not only be the responsibility of the ruling government, the rights and actions of citizens can be strongly supportive to such a cause. Urban planners can integrate the process of sanction of building plan with a compulsory clause of planting two appropriate trees in each house. Such a process has already been adopted in some cities but it's a long way to make it mandatory in each of the states of India.

4.0 DISCUSSION

Green City concept is an intelligent way of integrating urban planning with concern for environment. However, green city concept needs not only a scientific planning but also post monitoring and legal support. Mumbai has a Tree authority under Municipal Corporation, Hyderabad has support of Water, Land and Trees Act 2002 but 'Conservation of Trees in Non Forest Area' Bill in West Bengal is yet to get a nod. While most of the Indian cities suffer from dust and other gaseous air pollution, the greening of cities could provide an effective prevention against impact on health. Additionally, as has been seen in Calcutta, with urban greening a significant shelter for indicators of biodiversity can be created viz., birds and butterflies. Urban green can also help in controlling microclimate and recharging of ground water. And above all it can have immense bio-aesthetic value to augment quality of life.

REFERENCES

Benthall, A.P., 1946. The trees of Calcutta and its neighcourhood. Thacker Spink & Co. (1993) Ltd. W. Thacker & Co. London, 513. pp.

ENDEV, 2002. Report on Status of Calcutta's Parks and Gardens with an Integrated Management Plan. Society for Environment and Development, Calcutta.

Ghosh, A. K., 1988. Urban Ecology: A Case Study of Calcutta. Institute of Local Government and Urban studies. Govt. of West Bengal, 39 pp.

GoWB., 1990. Project Report for Greening Calcutta Metropolitan Area. Department of Forests, Government of West Bengal.

Randhawa, M. S., 1978. Beautifying cities of India. Indian Council of Agricultural Research, New Delhi, 15 pp.

CHAPTER XIX

Sustainability and Corporate Governance

1.0 HOW TO MEASURE SUSTAINABILITY?

A recent media article eulogized the Urban, Industrial and Commercial India since 1991 under the title of 'Rise and Shine' (S.L.Rao, The Telegraph, July 30th, 2007). The report reads eloquent on the apples from Fiji and China, Chocolates from the Switzerland, SUVs and Mercs adding glitter to the market place discovered in India. It claims 'The rich and the middle classes are very much better off' but remains silent on the subject of 30 percent Indian citizens living below poverty line with a sense of deprivation. The opening up of economy and abolition of license raj did not play magic to the millions. Any country which pledged for 'Sustainable Development' can never be termed as 'sincere' unless the common people also rise and shine.

What does sustainability really means? Indicators of sustainability have been debated upon over last 15 years. Indicators are supposed to provide information about physical social or economic system usually in numerical terms. Indicators can be used to describe the state of the system, to detect changes in it and to show cause and affect relationship. (Farell and Hart, 1998).

While monitoring indicators have been classically used they do not necessarily reflect the disparity along social stratification. Therefore Green National Accounting is perceived to indicate the quality of life in a country irrespective of rich and poor. Measuring Gross Domestic Product (GDP) became a part of System of National Accounts (SNA) developed by United Nations. SNA helps to understand a nation's economy, such as current account balance, wealth, and government income and expenditure. SNA helps at arriving policy decisions but tends to exclude important environmental factors- vital to understand the subject of sustainability.

Economists have now realized that unless a system of 'finite resource' management comes to place, development can never be sustained. Index for Sustainable Economic Welfare (ISEW) and the related Genuine Progress Index (GPI) are new tools to measure welfare which in turn indicates the trends of sustainability (Mottat, 1991). Adjusting income inequality, cleaning up pollution, preventing environmental degradation can help to arrive at such a process. Add to this the concept of 'ecological footprint focusing consumption and the flow of materials and energy through human system (Rees & Wackerhagel, 1996).

2.0 CORPORATE GOVERNANCE & SUSTAINABLE DEVELOPMENT

In the World Summit on Sustainable Development (WSSD) held at Johannesburg South Africa in 2002, 67 pages of implementation binding all countries to specific targets, deadlines and resource contribution to achieve Sustainable Development was announced. It focused amongst others on the need for change in unsustainable pattern of consumption and production. The responsibility of private sector becomes critical in such a scenario. Spelling out what private business social and environmental responsibilities, are should be essential. How much has been done?

An analysis of Indian Business Houses can help to understand the current level of efforts. While profit line remains vital life-sign, the increasing interest and involvement in achieving a better environmental management seems to be a welcome change. For example, International Certification (ISO 9000; 14,000; 18, 000) on general management and environmental system are increasingly gaining ground. Such a process is likely to lead to concerted efforts in reducing material use, water use and energy use and instill a doctrine of recycle and reuse. While such efforts in the long run benefits the corporate sector not only in image build up, and saving on expenditure, but also it helps the neighboring community to have a better quality of life.

But the corporate governance can never be restricted within the premises of its operational area. To ensure the doctrine of sustainable development the corporate sector needs to look beyond to the society at large. Social development projects initiated by corporate sector, as has been witnessed can prevent misuse of natural resource, cut down on health care expenses preventing disease.

The major stakeholders in the corporate governance are not only its employees, customers and suppliers, but also local communities and societies at large. Corporate sustainability therefore should not be confined to Economic performance but also to the Environmental and Social Performance.

Social Performance include assuring equity in employment, secure work environment with occupational safety, in service capacity building for future development; all these are part of social performance related to employees' welfare. But beyond the direct operational environment, it need to extend to the local community, when a company allocates resource and extent support for local community development irrespective of its turnover and profit after tax, Corporate care can be extended to development of Civic amenities, Health and Hygiene, Education, Vocational training for employment and income generation, Welfare and recreational activities and Environmental management.

Some corporate sector [e.g. Tata Steel] enterprises have introduced Corporate Citizenship Index (CCI) focusing on three key elements viz. Health (40%), Civic amenities (40%) and Environment (20%). Application of such or similar evaluation system is likely to reflect organizational performance against respective targets.

Obviously corporate social responsibility does not end at working environment and local community but extends beyond to the society at large. The involvement may be in the form of setting up Educational, Medical and Technical & Scientific Institutions imparting high level of training, and services of promoting research and development. Corporate contribution to R& D sector in India still remains very poor; as a result major burden is carried by the Government i.e. from the taxation revenue from people of India. Had Corporate sector substantially shared such responsibility, the allocation could have been diverted elsewhere.

Contributing to the government exchecker in an open and transparent manner can make the profit better. Direct involvement in

National and International cause e.g. disaster management, change of technology to reduce carbon emission also can make significant difference contributing towards sustainability.

'Sustainability' requires implementation of good environment management system. Investing in environment in the ultimate analysis helps the corporate sector in achieving better performance and assure better profit margin.

Environmental management involves 'Strategy Compliance' but calls for paradigm shift in other vital areas. One such being 'Resource Management'. Here resources mean raw material, water, land and energy (Besides capital and labour).

Land, Water and Energy sources remain finite resource (energy sources as on today also remain finite, while renewable energy may promise a better tomorrow, investment in renewable energy from corporate sector unfortunately remains meager). Growth in a finite environment calls for a serious introspection of current management system. Land management demands pollution free soil, prevention of erosion and green cover to ensure future productivity. For example old practice of leaving behind the mined area in a highly degraded condition has been changed by a government law and mine reclamation and closure plan has become mandatory before any mining disease is issued. Heavy use of chemical fertilizer and pesticides make it extremely difficult to introduce organic farming because of land pollution.

Water resource likewise remains also finite. In India, detailed accounting of ground water and surface water has been made. Already overuse and misuse of this vital resource has put "Red Alert" on Water Balance Sheet. Corporate sector, especially power and heavy industries, (besides Agriculture and Horticulture based industries) are to review the current practice of water use and follow a general principle to attend sustainability. These include:

- Water Conservation (through Check Dam, Rain water harvesting)
- Water reuse/ recycle (through appropriate technology adoption)

India being predominantly dependent on carbon based energy, due concern for energy management should be a cardinal principle of corporate governance. Industrial houses in India has started taking initiative for energy audit, use of product gases to cut down energy and take up alternative fuel sources like biodiesel, baggasse, rice-husk (at different level of demand). With additional incentive to gain from Carbon trading under Kyoto Protocol, industries are slowly gearing up to cut down carbon emission by adopting Clean Development Mechanism (CDM). This trend if accelerated can positively contribute to sustainability.

Environmental Management also demands Waste Management. Waste as a byproduct of production has been recognized as a major source of pollution and contamination of natural resource base as also a threat to Public Health. Waste Management system now ideally goes for

- Minimum waste generation
- Recycle and recover material from waste
- Disposal in an environment friendly manner

To achieve these objectives every corporate house must undertake exercises to

- Inventories & Categorise Waste
- Search appropriate method for reuse and recycle
- Develop waste management system in conditions where waste generation can not be prevented

Technologies are now available to manage solid waste, effluents and gaseous waste and generate wealth from waste. Examples of using Blast furnace slag for Cement manufacture, Fly ash for brick making, Coke Oven Gas & Blast Furnace Gas for power generation can be cited.

Sustainable Development stresses upon pollution free environment. Such an environment offers better quality of life and better productivity. As such pollution management, both water, air and noise demand serious attention; application of appropriate technology and allocation of fund should be a part of internalized system.

3.0 DISCUSSION

Corporate sector is expected to contribute significantly not only to the developing economic scenario but also to the social development of the country with a profound commitment to the philosophy of "Sustainable Development".

The current level of information, however inadequate it may seem, can provide the trend of investment in CSR in India. From a list of 35 corporate houses, where CSR activities have been documented by Centre for Social Market (CSM, 2006) an analysis shows that 50% of the respondents apparently failed to provide any data on Net Profit after tax for the given financial year or percentage of net profit spent on CSR. Further analysis of remaining 50% of respondents indicate that even when the net profit after taxation has been stated, amount spent on CSR has been spelt only by 11 corporate houses; it shows an average trend of investing 0.5 – 0.6% of net profit on CSR, sometimes going upto 7% (8% Titan) or even upto 9.7% but the later houses have apparently mixed up the amount spent on internal, staff related social responsibilities and community related programmes (e.g. Hindustan Lever). [Table 19.1]

Table 19.1 : Programme under CSR in selected Indian Companies

S.No.	Company Name	CSR Components
1	2	3
1	Asian Paints	Rainwater harvesting in housing society (N.A.)
2	ABB India Ltd.	Elementary Education (0.6%)
3	Aditya Birla Group	Healthcare, Education, Agriculture, Watershed development, Infrastructure (N.A.)
4	Ambuja Cement	Education, Health and Sanitation, Watershed Development, HIV Aids (N.A.)
5	Bajaj Auto	Trust works in 24 villages around Bajaj Plants (N.A.)

1	2	3
6	Ballalpur Industries	HealthCare, Primary Education, Natural resource Management, Capacity Building (N.A.)
7	British Gas India	Earthquake Rehabilitation, Education, Natural Resource Conservation (0.5%)
8	Cadbury India	Women Empowerment, Education and Health (N.A.)
9	Chambal Fertilizer	Agricultural Extension Service, e Education for farmers, Animal Husbandry, Vermicomposting, Herbal Medicine (2.5%)
10	Citi Group	NGO Partnership in women's micro credit (N.A.)
11	Colgate - Palmolive	Oral HealthCare in 50000 school, 26 million children, Education (N.A.)
12	Cummins India	Foundation led School in wheels Programme, Water Management in Village, Engineering College for girls in association with IIMA (N.A.)
13	FAB India	Girls Education (N.A.)
14	Forbes Marshall	Health and Education (N.A.)
15	Ford India	Health, Education, Disaster Relief, Environment (0.1%)
16	Glaxo Smith and Kline	Partnership with NGO'S for healthcare, Water Management, HIV / Aids awareness (N.A.)
17	Hindalco	Integrated Rural Poverty Alleviation Programme in most backward districts of four states (N.A.)
18	Hindustan Lever	Health, Environment and Education, both for internal and external CSR (19.7%)

1	2	3
19	Hero Honda	Healthcare, Education, Employment Generation Programme (0.5%)
20	I O L	Supports Conservation of Heritage Sites through its foundation in collaboration with ASI (N.A.)
21	Infosys	Education, Poverty alleviation, Women's Vocational Training, Health Care (1.5%)
22	Intel India	Education, Solid Waste Management, Environment (N.A.)
23	Jet Airways	Partnership with NGO "*Save The Children*" so far contributed Rs.2.5 million (N.A.)
24	Kanoria Chemicals	Social Development, Health & Education, Agriculture, Water Resource with NGO partnership (N.A.)
25	Lupin	Rural Welfare benefiting 1900 villages in 11 districts in four Indian states (N.A.)
26	Mahindra & Mahindra	K.C. Mahindra trust promotes education through scholarships, interest free loans and setting up chair in Universities (N.A.)
27	Nestle	Partnership with NGO'S in community projects on Education, Health, Welfare and Infrastructure Development (N.A.)
28	Nokia	Use of recycle paper for packaging; Working with NGO's for disabled children (N.A.)
29	Phillips India	Social Audits of Suppliers (N.A.)
30	Phoenix Yule	Education, Health, Environment (0.6%)

1	2	3
31	TCS	Adult Literacy Programme (N.A.)
32	TATA Steel	Integrated Rural Development, Reproductive Healthcare, Education, HIV Aids Prevention, Disaster Relief, Sports Training (5.2%)
33	Titan	Empowerment of women in Karnataka, Child Care, Infrastructure Development in partnership with NGO's (7%)
34	Wockhardt	HIV Aids, Education and Research Foundation in collaboration with Harvard Medical International (N.A.)
35	Wipro	Not Specified (0.003%)

Table 19.1.1: Case Studies on CSR from India

S.N.	Name of Company	Net Annual Profit #	% spent in CSR	CSR Components
1	2	3	4	5
1	ABB India Ltd	150.0	0.9	Elementary Education
2	Aditya Birla Group	1228.0	N.A.	Healthcare, Education, Agriculture, Watershed development, Infra-structure
3	Ambuja Cement	222.0	N.A.	Education, Health and Sanitation, Watershed Development, HIV Aids
4	Ballalpur Industries	2244.0	N.A.	HealthCare, Primary Education, Natural resource Management, Capacity Building

1	2	3	4	5
5	British Gas India	11837.0	0.5	Earthquake Rehabilitation, Education, Natural Resource Conservation
6	Chambal Fertilizer	89.4	2.5	Agricultural Extension Service, e Education for farmers, Animal Husbandry, Vermicomposting, Herbal Medicine
7	Ford India	493.0	0.1	Health, Education, Disaster Relief, Environment
8	Hindustan Lever	125.5	19.7	Health, Environment and Education, both for internal and external CSR
9	Hero Honda	846.0	0.5	Healthcare, Education, Employment Generation Programme
10	Infosys	1243.0	1.5	Education, Poverty alleviation, Women's Vocational Training, Health Care
11	Intel India	145.0	N.A.	Education, Solid Waste Management, Environment
12	Kanoria Chemicals	22.66	N.A.	Social Development, Health & Education, Agriculture, Water Resource with NGO partnership
13	Phoenix Yule	511.5	0.6	Education, Health, Environment

1	2	3	4	5
14	TATA Steel	3604.0	5.2	Integrated Rural Development, Reproductive Healthcare, Education, HIV Aids Prevention, Disaster Relief, Sports Training
15	Titan	24.95	7.0	Empowerment of women in Karnataka, Child Care, Infrastructure Development in partnership with NGO's
16	Wipro	9539	0.003	Not Specified

In Crores

As expected, most of the Corporate Houses largely show interest in providing services towards Education and Health Care, followed by programme related to empowerment of Women; very few houses made special contribution to combat HIV / AIDS; similarly not many organizations have involvement in promoting 'Rainwater Harvesting' or 'Renewable Energy': Unless priorities are fixed to contribute towards 'Sustainable Development' process, the focus of CSR may remain rather diffused.

The classical concept of economic development as an one dimensional thought involving capital and labour has changed with the publication of "Club of Rome's report on 'Limits of Growth' in 1970's and 'Our Common Future' prepared by Bruntlund Commission in 1980's.

The World Conservation Strategy of 1980 laid the pathway which culminated in the adoption of 'Agenda 21' at Rio Conference on Environment and Development in 1992. Agenda 21 identified the major stakeholders in the process of attaining Sustainable Development and UN Commission on Sustainable Development was set up in 1993. The World Summit on Sustainable Development at Johannesburg in 2002 once again called for commitment of the member nations. Millennium Development Goals (MDG) laid stipulation to reduce poverty by half by 2015 as poverty and population act as major hurdles in developing nations while consumerism posses biggest challenge to sustainability in the developed nations.

The norms set up by UN for Responsibilities of Transnational Corporations and other business [UN Doc.ECN 4/Sub2/2003/12/ Rev2(2003)] provide a set of human rights obligations directed at companies; such obligations, it is expected, can be improved by way of domestic laws of individual states (Kinley, 2007). It is believed to be a part of CSR initiative but the states may not be 'always willing or capable of strengthening human rights regulations'. Whether such responsibilities should best be voluntarily shared or imposed by state remains an open question. In broader sense, CSR does not only include Human Rights, it goes beyond the listed rights and looks a broader social responsibilities depending on local situation.

Corporate Sector is expected to change their mode of thought process and recognize 'Social capital' as an important lever in Economic Development (Rainey et. al, 2003). Periodic social audit may help to assess the extent of fulfillment of such social commitment (Tata Steel, 2005).

REFERENCES

CSM, 2006. Who's Who in CSR in India. A Resource Book. Centre for Social Markets, Kolkata

Farell, Alex and Hart, Maureen, 1998. What does Sustainability really mean? The search for useful indicators. *Environment*, 40(9): 1-12

Kinley, David. 2007. CSR Politics: reflections on the UN "Norms". *The Contemporary Manager*, 12(1): 43 – 48.

Moffat, I. 1991. On Measuring Sustainable Development Indicators. *Int. J. Sustainable Development and World Ecology*, 1(2): 97.

Rainey, Daniel, V., Robinson, Kenneth, L. Allen, Ivye and Christy, Ralph, D. 2003. Essential Forms of Capital for Sustainable Community Development. *Amer. Jour. Agric. Econ*, 85(3): 708-715.

Tata Steel Limited, 2005. 3rd Social Audit, 1991-2001.

Wackernagel, M and Rees, N. 1996. Our Ecological Footprint: Reducing Human Impact on the Earth. Gabriola Island B.C: New Society Publishers.

CHAPTER XX

Protecting the Environment: Three Decades of Indian Progress

1.0 GENESIS

The environmental concerns in India emerged only after the UN Conference on Human Environment and Development held at Stockholm, Sweden in 1972. India was represented by the then Prime Minister Indira Gandhi; Gandhi said 'we do not wish to impoverish the environment any further, and yet we cannot for a moment forget the grim poverty of large numbers of people. Are not poverty and need the greatest polluters?' That was a time when India, liberated from colonial rule was a young nation of 25 years old. Indira Gandhi linked poverty with pollution keeping India's perspective. However the industrialized nations were concerned about pollution not related to poverty but due to the sheer process of industrialization with new technologies, capital and labour. Gandhi was right when she pleaded helplessness in saving wildlife and forest in a society so much dependent on free natural resources in an environment of unemployment, lack of access to basic living standards and deprivation at all levels. One has to remember the global concern for planet earth expressed in the fastest developing economy of USA in the form of observation of Earth Day on April 22, 1970.

India has passed through a long passage of time since then along with rest of the world. While Gandhi pleaded her inability to protect wildlife from hungry, poor millions, the Wildlife Protection Act was passed in the same year of 1972. It is regarded the first attempt to protect in a comprehensive manner, the wild fauna but remained faulted at birth by omitting threatened, endangered and vulnerable species of flora – an integral part of wildlife scenario.

The year 1972 was also significant at an international level. In an attempt to prepare a mathematical model for world system, back in 1968 thirty individuals from ten countries diverse in background and interests assembled in the City of Rome, being concerned with 'the present and future predicament of man'. A 'Club of Rome' was formed which undertook "to faster understanding of the varied but interdependent components – economic, political, natural and social – that make up the global system in which we live, to bring that new understanding to the attention of policy makers and the public worldwide". The result was publication of a document called 'The Limits to Growth' in 1972 by Prof. Dennis Meadous and his team at M. I. T., Boston, USA. The highly publicized projections through mathematical simulation, indicate that 'continued growth leads inevitably to a catastrophic collapse of the world ecological and economic system" (Baumet and Oates, 1979). It perhaps reminds one of the Malthusian prediction that population without check would overrun supply of food with misery, pestilence and famine. The parameters used by the Club of Rome model included 'Resource', 'Population', 'Food per capita', 'Pollution' and 'Industrial output per capita'. Living in a finite space, resources and capacity to absorb pollution will inevitably lead to collapse of the world system in a declining path of development.

Back in India, National Committee on Environmental Planning and Coordination (NCEPC) was formed by the Government and a series of events followed rather fast to keep with global demand for environment management. In the inaugural programme of NCEPC at New Delhi on 12th April 1972, the Prime Minister of India re-emphasized the need 'to bring basic amenities within the reach of our people....without alienating them from nature and environment' and argued to move away 'from the single dimension model which equates growth of GNP with development'. It was apprehended 'as our development progresses perhaps we also shall have environmental problems akin to those in the developed countries...... (but) concern with economic and social development need not be a choice between poverty and pollution'.

An analysis of the progress of India in the path of integrating environment with development has been made in this essay taking into consideration the key parameters for evaluation.

2.0 POPULATION CONTROL

'It is an ever-simplification to blame all the world's problem on increasing population' said Indira Gandhi while addressing the UN Conference on Human Environment and Development in 1972. Nevertheless, it remains a fact that growth in a finite environment could be more feasible with a manageable population to avert alarming change of Human : Land, Human : Water or Human : Forest ratio. Per capita availability of all major natural resources have become less than a third of what was available in 1947. India's geographical area remaining static and economic policy from the 1st to 8th Five Year Plan period being conservative, the developmental goals faced crisis of shortage of raw material and capital, besides lack of access to modern technology.

The family planning programme to control population did not show any spectacular results. India has crossed 1 billion mark by the turn of the last century. With opening up of market in a globalized world, India's economy is now said to be "buoyant". But the population growth rate still remains alarmingly high, even though morbidity and mortality rates show downward trend. India remains imprisoned within an unresolved crisis of human growth rate.

The recent "National Population Policy" (NPP) released in February 2000, paid special attention to the health and education of women and children to achieve population stabilization for the country by 2045. This suggests a paradigm shift to reproductive and child health care from the 50 years trial of achieving control through vasectomy and tubectomy and a series of incentives; although 97 per cent of sterilization are tubectomy in recent years, Government did not consider it rational to promote vasectomy in male, making a glaring gender bias. 'Despite the promising performance so far, it seems unlikely that India will attain the goal of a replacement level fertility of 2.1 in 2010 as was projected in NPP 2000' says Pravin Visara (2002). Unless average growth rate of 1.9 (2001) is reduced, India will continue to face a crisis of plenty of people. The census figure of 1951 (361.0m) and 2001 (1027m) makes one ponder over the future.

The National Conservation Strategy and Policy Statement on Environment and Development, 1992 puts 'Population Control' as the number one item in the list of priorities and strategies for action. The National Environment Policy, 2006, surprisingly remains silent on the issue of 'Population Control' while highlighting key environmental challenges. It also makes a causal reference to NPP.

3.0 CONSERVATION OF NATURAL RESOURCES

3.1 Land Resource

The geographical area of India covers only 2.4 per cent of global space but contains more than 16 per cent of human population and largest number of livestock in the world. The actual reporting area for land resource is pegged at nearly 305 million ha., out of 329 mha., so that utilizable resource becomes more restricted. Per capita land availability has been reduced to less than a third than what was available in 1947. As has been pointed out elsewhere, the land available for non-agricultural use is only 6.8 per cent, the rest being largely used for agriculture barring 5.8 per cent barren or uncultivable land. India faces therefore a serious crisis with land resource.

The crisis in the management of land resource has been repeatedly recognized by the National Agricultural Commission, Land Use Boards of the State and the Ministry of Agriculture; the Environment and Development Policy Statement of 1992 by the Ministry of Environment has recognized Land resource demanding serious attention. It acknowledged "of about 329 million hectare, 175 million hectares of land require special treatment to restore such land to productive and profitable use. The degradation is caused by water and wind erosion (150 mha.), salinity and alkalinity (8 mha.) and river action and other factors (7 mha.)". The same policy statement mentioned the steps so far taken for reclamation viz., survey by the All India Land Use and Soil Survey Organization, Catchment area treatment in selected river valley project, Drought prone area programme, Desert Development programme, Assistance to Control Shifting cultivation etc. None of these helped to turn the table. The policy called for an integrated land and water management approach, amelioration of water logged and salt affected lands, prevention of land fragmentation, protection of good agricultural land etc. (MoEF, 1992).

A look at the present position may be revealing on the status of land resource in 21st Century India. A report on Status of Environment in India by the Centre for Science and Environment, New Delhi (1995) recorded nearly 130 mha., of land as wasteland, including degraded forest land (93.66 mha.), saline alkaline area, wind and water eroded area and degraded forest land (35.85 mha.). If one compares the figure of 1992 paper, i.e., 175 mha., of land under degradation with that of 1999 figure of 130 mha., it can be presumed that 45 mha., has been reclaimed between 1992-1999, i.e., at the rate of more than 6 mha., per year! In reality government statistic puts the figure at 1.5 mha., per year i.e., in 7 years at most 10.5 mha., could have been reclaimed. Authentic data on reclamation of degraded land is an urgent necessity.

What is however most disturbing is the lack of a definite Policy on Land Management evenafter nearly six decades of independence. Excellent data on land capability and limitation generated by Government of India organization will not help unless a positive policy backed by appropriate legal instrument (Act & Rules) become available. The National Environment Policy, 2006 only acknowledged the land degradation due to proximate causes but does not call for a National Land Policy and legislation to curb the loss of this finite resource in a developing economy.

3.2 Water Resource

Its not only the agricultural policy, other vital sectors of development like "Water Resource", has raised many questions in retrospect. The strong arguments in favour of western model of taming the rivers in the name of multipurpose river valley project fell flat in terms of real environmental and social cost and benefit. The total virgin forest area sacrificed due to submergence for creating of reservoirs, since 1947 has never been quantified neither the loss of unchartered genetic resources lost forever. Only in 1980's, the data collected by agricultural scientists on wild rice varieties (each with unique traits) from Kerala's Silent Valley showed the folly of such projects (Khoshoo, 1986). The social cost of displacement of 30 million nameless, faceless rural Indians remain in the dark. Records of evictions leave aside any compensation are non-existent in most of the cases. No national policy of 'Rehabilitation and Resettlement of Project Affected Families' was announced till 2005. India is now recognized as the 3rd largest dam builder in the world. The large dam building panorama from Tehri to Narmada have attracted widespread media attentions; pro and anti dam arguments were never seen from an unbiased view point and but at the end (lasting a period of more than 25 years) pro-dam lobby won over due to continued policy of the Government of India. In the interest of Gujarat, the people of Madhya Pradesh (MP) and Maharashtra suffered. Malwa and Nimad area of MP was never a backward area. Historical evidences record at least 12 different empires and various civilizations flourished in Malwa and Nimad over last 7000-8000 years. Politicians declared it as a backward area and 'the remedy for all backwardness was to build dams on the river'. The Naramada Project was cleared by the Government of India on April 13, 1987 but the controversy continues. Medha Patekar led Narmada Bachao Andolan talked of social cost of evicting 2,00,000 people; hardly anyone pointed out the case of submergence of vast tract of virgin forests in the era of biodiversity conservation. Eviction not only destroys settlements but uproots the system of craft and culture. When it was pointed out in a recent conference on "Folklore" the specialists paused. No research has ever been seriously undertaken. More than half of the river valley projects in India remain incomplete in 2006 at an incalculable time and cost overrun.

The water rich India however suffers from inequitable distribution. Without environmental planning the latest project on 'Interlinking of Rivers at a cost of Rs. 5,60,000 crore is supposed to guarantee such equitable sharing. Debates raged over the country; finally Himalayan component of interlinking of rivers project is kept in animated suspension.

The lure of productivity had its tremendous toll on groundwater reserves of the country. The water balance is thrown to the wind multi cropping in a monoculture (Paddy – Paddy – Paddy)reined supreme. A wheat crop need six irrigation. What is way to avoid overdoing if the green revolution mantra was national chant for 30 years!

3.3 Atmosphere

The concern for atmosphere was first expressed in the form of Air (Prevention and Control of Pollution) Act, 1981, exactly a year after the Union Department of environment was formed. The preambler part of the act states 'whereas decisions were taken at the

United Nations Conference on the Human Environment held in Stockholm in June 1972, in which India participated, to take appropriate steps for the preservation of natural resources of the earth which, among other things, include the preservation of the quality of air and control of air pollution'. The said act is passed by the Parliament in the Thirty Second Year of the Republic of India. The act has been amended several times, standards for ambient air quality has been notified, sectoral standard for emissions from industrial, automobile, power, mining have all been fixed. Environment Protection Act, 1986 further facilitated to act upon polluting sector by empowering officials of 160 other organization, in addition to State Pollution Control Board, to enter and inspect any unit to detect malpractices. The Air Act had also laid down penalties (fiscal penalty and imprisonment) for contravention of the provision of the act. The Air act, under section 19, empowered the State Government 'declare .... any area or areas within the State as air pollution control area ... (and) subject to the provisions of this act no person without the previous consent of State Board, establish or operate any industrial plant in an air pollution control area'.

In 1992, Policy Statement stated 'Global atmospheric changes resulting in altered temperature and precipitation and rising sea level are no longer within the realm of mere theoretical possibilities'. It identified "Atmosphere" as a major area demanding action. Thrusts to be given on use of clean fuel, clean technologies, proper location of the project to minimize adverse impact on people, developing coping mechanism for future climate changes etc. A total of seven agenda items featured for "Atmosphere" in 1992 policy. The results of action taken through the Air Act, 1981 and Environment Protection Act, 1986 do not feature in the document.

Then came the National Environment policy, 2006. Under the section of Pollution Abatement, the subject of "Air Pollution" features in a general descriptive manner. Specific action to be taken, include energy conservation, improved fuel wood stove, monitoring and enforcement of emission standards, city based action plans, formulation of a national policy for urban transport, partial substitutions of fossil fuels by bio-fuels etc. A very generalized set of pious objectives without set target, quantified data on present status and set target for reduction!!

In the meanwhile, India has joined major international conventions related to changing Atmosphere. India is signatory to the Vienna Convention for the Protection of Ozone Layer, 1985 and Montreal Protocol on Substance that Deplete Ozone layer, 1987. India has taken significant steps to control CFCs, Halons and other ozone depleting substances but the small scale units did not get the benefit of fund used for replacement received from Multilateral Fund. While Ozone Depleting Substances (ODS) is being reduced, (over 1999-2010), the carbon emission factor along with other noxious gases remains a major problem. As a member of UN Framework Convention on Climate Change, 1992 and signatory to Kyoto Protocol, 1997, India is obliged to take necessary action. Green House Gases (GHG) emission from India is only 3 per cent of global emission. In per capita tonnes, the National Environment Policy, 2006 states 'it is 23 per cent of the global average, 4 per cent of USA' per capita emission in 1994.

The Policy paper does not however present the figures of decadal change. It mentions "Since GHG emissions are directly related to economic activity, India's economic growth will necessarily involve in GHG emission from the current extremely low levels. Any constraints on the emission of GHG by India, whether direct by way of emission target, or indirect, will reduce growth rate". A surprising statement sounding like the statement of US President, who withdraw from Kyoto Protocol on the pretext that why India and China have been exempted from being given a target date for emission reduction. That India emitted a little over 0.4 tonnes CO_2 equivalent per 1000 US dollars in 2002 does not make the country immune from its commitment to abide by Kyoto Protocol's guiding principles.

In the domestic front, only capital city of Delhi changed over to CNG fuel in Public transport system. Commercial vehicles over 30 years old continue to ply on the roads of Kolkata. Conversion of Auto rickshaw from Diesel to LPG has become a mockery in the same city. Unlimited growth of vehicle population without commensurate increase in road space kept India's air quality very critical. Add to this, violation of Air Act by the industrial sector like Sponge Iron and Ferro Alloy factories; with impunity. Air Pollution continues to be one of the major killers in India after 60 years of Independence. Keeping in view the monitored data available on air pollution, the Supreme Court in its various judgments have identified 16 major cities as equal to or more pollution to Delhi (MoEF, Annual Report, 2005-2006); the list includes Mumbai, Chennai, Kolkata besides Hyderabad, Bangalore, Pune – the emerging economic hubs. Compulsory switch over to CNG / LPG in Ahmedabad, Kolkata, Pune, Kanpur was planned in 2002 but still awaits implementation. CPCB had earlier identified more than 100 areas of critical air pollution [CPCB had also identified 1551 large and medium industries in 17 categories of highly polluting industries; they were given time schedule to install necessary pollution control equipments; at least 1351 units have installed the facilities, 178 units have been closed down, 22 units were reported to be defaulting in 2002-03. However, pollution profile of major Indian cities did not change to any significance].

The Auto Fuel Policy and Capacity Building for Industrial Pollution Management is yet to provide much desired results.

3.4 Biodiversity

India can be ranked as one of 17 countries of mega diversity. With only 2.4 per cent of global space it holds nearly 8 per cent of recorded biodiversity from the world. The diversity of ecosystems and species is well documented and the works on genetic diversity is in progress. India is one of the few countries to have enacted a Biological Diversity Act in 2002 and Rules in 2004. With this impressive profile India's biodiversity can be expected to be safe and secured.

However, the 'National Biodiversity Strategy and Action Plan' (NBSAP) done at a cost of Rs. Two crores still remain awaiting to be accepted. The extensive process of consultations ultimately bears to fruit of action. The system of governance reigns supreme against the will of the civil society.

Arguments may be put forth on the long series of recommendations, some highly radical that were not liked by the government; but there could always be a system of open debate and negotiation rather than mute silence mode. The document in draft form was never cleared by the Ministry of Environment and Forests (MoEF).

Questions are also logically raised on the practicality of implementation of the Biological Diversity Act with regard to formation of Biodiversity Management Committee (BMC) at the local level; in 2,50,000 local Panchayat level institutions, such a process can expectedly bring an accountability of resource use and benefit sharing mechanism. Since the rules have been framed in 2004, the rate of progress of constituting State Biodiversity Board (SBB) can be first indicator of progress. By the end of 2006, at least 50 per cent of the States and Union Territories are yet to form SBB, leave aside BMCs. To form BMCs within a time frame linkage between State Environment and Panchayat Departments is essential; unfortunately effective dialogues are still lacking.

Even if the BMCs are formed, who, when and how will provide 'People's Biodiversity Registers' (PBRs)? India has witnessed an alarming rate of decline of taxonomists during post 'Convention on Biological Diversity' (CBD) era, when it was most needed. Indian Universities relegated "Taxonomy" to the low priority area and funding agencies have little to offer research support. The largest employers of the taxonomists in the country (BSI & ZSI) are facing job-cuts or inappropriate recruitment policy putting non-taxonomists in the positions meant for taxonomists.

This mismatch between the professed objective and actual is now all too apparent. But no efforts seem to have been taken urgently to correct the position. Large repositories of taxonomic collections are facing crisis of maintenance; once lost, incalculable damage to science and irreplaceable loss of treasure will be stared at the scientific community. Remedial measures have been suggested in the past and again in the approach paper for 11th Five Year Plan. One tends to suffer from suspicion in action because of past experience.

While India can boast of having 600 odd protected areas and several ex-situ conservation efforts, much remains to be documented starting from microbial diversity in the soils of different agro-climatic zones to the marine life in the deep sea of India's exclusive economic zone.

Policy, Acts, Laws and Institutions are there, the need of the hour is implementation with a clear vision, and positive support from the Government of India, as India is deemed a 'Country of Ultimate Responsibility' by IUCN. One only hope that is discharged honestly. But managing marine protected area with cadre service officers of Indian Forest is but one instance of putting square peg in a round hole.

4.0 DEVELOPMENT POLICIES

3.5 Agriculture

Having said very prophetic words, the path chosen for development, sadly was influenced by the powers of a polarized world. In the field of agriculture, India wanted to be self sufficient and the roadmap of green revolution brought in the essential inputs of hybrid seeds, chemical pesticides and fertilizers, big dams and canal irrigation. In a labour intensive agriculture the indigenous technologies of water harvesting, use of farmers' varieties of seeds, organic fertilizers and pesticides were relegated to the backyard. The cost was enormous.

West Bengal, the homeland of rice field, had 5000 varieties of indigenous seeds in 1947 – not more than 500 could be located now in the farmers' field. Valuable genetic traits of each of these varieties are lost. Over use of irrigation water led to salinization and water logging in many parts of India. The debt traps of farmers become stronger, leading to suicides of thousand agriculturists from Andhra Pradesh to Maharashtra.

Swaminathan (2004) in his Anniversary Lecture to Shastri Indo – Canadian Institute, remarked that 'the fatigue of the Green Revolution' occurred due to 'marketing factors, meteorological factors, that affect variation and management factors – such as management of water and land'. He admitted the ecological problems of over irrigation, (water logging and salinization) and over use of pesticides (leading to breakdown of resistance to pesticides). Swaminathan recalled his lecture delivered in 1968 warning farmers 'not to over do anything', which he said was directed to the Government. As opposed to area expansion in agriculture between 1950-1966, increase in productivity was said to be the target of green revolution. Whatever may be the reason and defensive arguments proffered, the end result is loss of soil fertility, disbalance in biological cycle, pesticide contamination, exhaustion of groundwater and above all loss of accessibility to a large number of indigenous farmers' varieties. The new perception is 'ever green revolution noted in the principles of ecology, economics and social and gender equity (Swaminathan, 1996). One only wonders how it has taken more than 30 years to come up with this paradigm shift of thought.

The problem of pesticide pollution of our rivers and wetlands, food, milk and water can be ascribed as side effects of the campaign to promote 'green revolution'. The state of Punjab, the fastest growing agricultural state has reported not only a sharp decline in productivity but also pesticides in mothers' milk meant for feeding new born babies; the increase in the number of incidences of blood cancer has been highest in the area of use of non-biodegradable persistent, chlorinated pesticides during the period of green revolution (*Down to Earth*, June 15, 2005).

3.6 Forestry

In case of forest resources, India had continued to follow the colonial policy of considering forest timber as a major good for earning revenue. Commercial clear felling and conversion of forest for non-forestry purposes was rampant till the announcement of the Forest (Conservation) Act in 1980. In 1952, India's national Forest Policy aimed at covering 33 per cent of land area. The Prime Minister Indira Gandhi while inaugurating the XVIII Meeting of Central Board of Forestry on 25th August 1980 admitted failure of governance. "The ruthless and indiscriminate felling of trees, especially on our Himalayan slopes for immediate profit either of contractors or in the name of development has proved hazardous", she had noted. The revised 20-point programme of the then government refers to the need to pursue vigorously programme of afforestation. Clear felling of forest by the State Forest Corporation did incalculable harm by 1980. The rate of conversion for non-forestry was alarming. The Forest (Conservation) Act changed the perception of 'forest is free for all'. It called for clearance from authority before any conversion of forestland for non forestry purposes and demand compensatory afforestation. Needless to say the genetic variability of prime virgin forest can never be recreated by human society. The act had slowed down the rate of conversion till 2000 but in the high wind of globalization quicked the forest clearance. In a haste to provide economic boost, prime forest areas in iron-ore rich forest of Chattisgarh and Jharkhand are given off; as a matter of fact 45 per cent of total forest converted 8,47,000 ha., between 1980-2003 was during 1999-2003! Large scale forest clearance for mining is supposed to be compensated by the miners by way of compensatory afforestation – that however can never recreate the genetic diversity and lost habitat for wildlife. But even compensatory afforestation clause is blatantly violated by the State Forest Departments. A recent CAG report (2005-2006) hauled up the offending state governments. Earlier Supreme Cóurt of India had issued strictures and guidelines for future utilization of deposited fund. Sadly, till date there is no comprehensive monitoring mechanism to account for lost forest vis-à-vis forest recreated under compensatory afforestation clause. Supreme Court's order to form a "Compensatory Afforestation Management and Planning Authority" (CAMPA) in 2002 and valuation system based on "Net Present Value" is yet to be bear evidence of a reversal of forest policy. States continue to fail to deposit the money to the fund and even irregularly transferred the money for other purposes, CAG report pointed out (Ghosh, 2007).

The other laudable policy change witnessed was in 1988 with Government support for Joint Forest Management (JFM) concept. Nearly two decades have passed but the excellent concept of JFM in many states is facing an ever increasing threat of failure. The reasons primarily are attributed to lack of transparency of forest department controlled Forest Protection Committee (FPC) and lack of accountability. This was more clearly pointed out in a recent study on the impact of JFM on the livelihood of the people in South West Bengal (Banerjee, 2006).

The new National Environment Policy, 2006 acknowledged that 'disempowerment (of forest dwellers) has led to forests becoming open access in nature, leading to their gradual degradation in a classic manifestation of the Tragedy of Commons'. It only assures of an innovative strategy to increase forest cover to 33 per cent by 2012!

3.7 Industries

The industrial development of India during 1947 – 2006 has witnessed significant investment, transfer of technology and export orientation in post 1991 period. Heavy industries like Iron and Steel, Mining, exploration of Oil and Gas and linked activities in the Refineries and Petrochemical have all been areas of expanding activities. Indian industrial giants have acquired complexes across the world; automobile industries changed from a monopoly market to a wide range of selection for the buyers.

India's current GDP growth at 8.5 per cent owes much to the industrial sector.

Industrial pollution has been a matter of concern since enactment of Water Act in 1974, Air Act in 1981 and Environment Protection Act (EPA) in 1986. A.Pollution Abatement Policy was declared in 1992. Regulations, legislations, agreements, fiscal incentives and other measures like setting up industry specific standards have been major activities during last 30 odd years.

However, an analysis of the provisions of laws admitting pollution above permissible limit as a cognizable offence, shows that hardly any case of imprisonment clause has been used in India even after serious violation of Water or Air Act or EPA, 1986. Strict actions were called for but a policy of appeasement pervaded the system. The Government of India follows a policy of persuasion than prosecution and industries like Sponge Iron and Ferro Alloy sector in recent times continue to take the advantage of this lax attitude.

Standards have been reviewed for specific industries but that really didn't help. For instance, in spite of Ambient Air Quality being notified more than 10 years ago, at least 65 urban centers in 23 States and Union Territories continue to violate the same. The State Pollution Control Board could not impose the standard especially with reference to Respirable Particulate Matter (RSPM) and Suspended Particulate Matter (SPM), often also for Sulphur di-oxide (SO_2). Motor vehicles have been identified as major source of air pollution (MoEF Annual Report, 2004-05, p-50) but no sincere attempt can be seen even to replace age-old public vehicles or to more environment friendly fuel in majority of the urban centers of India. Strangely, the pollution under control certificate is only called from private owners.

Recently a total of 2301 medium and large scale polluting units under 17 categories were identified in 27 states of India. Central Pollution Control Board (CPCB) claimed that 1927 have since complied and 235 have been closed down while 139 units are marked "Defaulting". No significant change in air quality however has been reported.

A National Environment Appellate Authority (NEAA) has been set up in 1997 but complaints of industry specific pollution e.g., Sponge Iron and Ferro Alloy units remain unattended.

On the better side, the Government has set up a system of providing Financial Assistance for setting up Common Effluent Treatment Plants (CETP) to the tune of 50 per cent of the cost as State and Central subsidy and 30 per cent as loan from financial institutions. But most of the Municipalities in India continue to discharge untreated effluents into the surface water system with impunity.

While no significant or perceptible change in India's pollution profile is yet visible, efforts cannot be undermined. One such being drawing up a 'Charter on Corporate Responsibility for Environmental Protection' (CREP). After a series of industry specific consultation CREP was adopted in March 2003 for 17 categories of polluting industries. Eight task forces have been formed to oversee the process of implementation. Results are yet to emerge.

A grant-in-aid scheme on development and promotion of clean technology was also initiated in 1994 with the objective of achieving the goals; since inception in 1994, activities undertaken include (I) 'Carrying Capacity Studies' in some identified areas like National Capital Region, Doon Valley, Damodar River Basin etc., (II) National Resource Accounting, (III) Life Cycle Assessment (LCA) studies in Steel, Thermal Power, Pulp and Paper, Cement industries. Many of the LCA study reports have been published, since field demonstration of innovative clean technologies have been made. Again no significant impacts of this decade long project is seen.

The issue of River Clean up Programme – which are interlinked with discharge of polluted industrial effluent demand a serious introspection. Although volume of domestic wastewater may exceed the volume of industrial effluent, the toxicity factor of the latter is known to much higher. This was revealed during Ganga Action Plan. Of the 340 projects sanctioned so far, 181 projects have reportedly been completed. The end results are yet to emerge. Citizens of India agree that Ganga pollution has not been reduced to any significant extent.

3.8 Mining and Quarrying

The mining sector has been identified as one of the major industrial activity area leading to possible air, water and noise pollution besides causing extensive land degradation. The huge overburden dumps created during opencast mining remain a major problem and rehabilitation of the abandoned mined area in a phased manner was one of the objective stressed in the Policy and Strategy statement on Environment and Development, 1992.

Restriction on mining and quarrying activities in the sensitive areas such as hill slopes areas of natural springs and areas rich in biological diversity was another professed objective.

However, any study of the old abandoned mined area will reveal that no effective measure has been taken so far. What however is noticeable is more stringent compliance of submission on environmental planning along with mining plan. Mine closure and reclamation plan is now stipulated with every new mining lease. Post 1988 period, a series of efforts of afforesting completed part of ongoing project has been seen. Import of new technologies like 'Ripper-dozer' in shallow mineral deposit areas (e.g., Bauxite in Maharashtra and Jharkhand) has shown impressive results.

But what is worrying the environmentalists is large scale iron ore mining in prime biodiversity rich forest areas of Jharkhand and Chattisgarh during post-2000 period. Increasing international demand of iron ore led to one of the fastest conversion of forest area into mining zone. To restrict high quality ore export, a recent decision has been taken (end 2006) but mining will continue.

Lack of effective monitoring and lack of any penal action for violation makes implementation of Environment Management Plan an unfinished agenda in many mining projects in wide areas of West Bengal, Bihar, Jharkhand, Chattisgarh, Madhya Pradesh, Goa and other parts.

Quarrying stone chips still remain a major critical unresolved issue. Most of stone quarries are operated by employing casual labour and occupational health hazard appears rampant. In West Bengal a Status Report on Environment (DoE, 1999) shows the extent of horror. The directives of Hon'ble Supreme Court, the report pointed out is openly violated by the stone mine operators in West Bengal – Bihar border.

A critical analysis will show that much of the professed objectives of 1992 Policy and Strategy remain unfulfilled. Strangely enough, the Annual Reports of the Ministry of Environment and Forests for the year 2002-03, 2003-04, 2004-05 and 2005-06 have no reference to any mining area rehabilitation programme under section 'Regeneration and Development'. The new National Environment Policy 2006 however continues to refer to ensuring 'environmental restoration after decommissioning of industries, in particular mine closure in all approvals of mining plans and institutionalize a system of post monitoring of such process'. No one can say when it will start.

3.9 Tourism

The focus on India as a tourist destination might have been sharper during last two decades. Tourism in India had been classically concentrated around historical sites (Rajasthan), sea beaches and mountains. Wildlife and Nature tourism gradually gained ground over last 30 years largely due to publicity connected with establishing more than 600 Protected Areas (Sanctuary, National Parks, Biosphere Reserves, Marine Reserves, Tiger Reserves, Elephant Reserves etc.). With growing volume of tourist traffic, environmental concerns were expressed in Policy and Strategy Statement of 1992. It stated that 'to ensure sustainable growth of tourism without causing irreversible damage to the natural environment, activities related to tourism should take care of (I) Carrying capacity and support facilities, (II) Tourism in harmony with local people, (III) Regulatory control of tourism in sensitive areas such as hill slopes, islands, coastal stretches, National Parks and Sanctuaries.

Consequently, the term 'ecotourism' became a key word. Ecologically sustainable tourism can only be promoted through a process of understanding carrying capacity of the area and guiding the tourists to follow a set of norms. However, all hill stations famous for tourism have witnessed totally unsustainable growth violating norms of building rules, overtaxing limited water supply and sanitation system, jeopardizing capacity of solid waste management, which in any case is largely lacking.

Any visitor to Shimla in Northwest to Darjeeling in eastern Himalaya will vouch for the ever deteriorating civic scenario. The local resident population face unprecedented shortage of water supply in every hill station, garbage piled up are dumped down hill slopes, noise pollution level soars high. The visitors to the sea side, forests and hills are now being restricted with regard to use of plastics. But no monitoring on ban on plastic seems to be in place.

The shortage of infrastructural facility and tourism beyond carrying capacity remains a national syndrome. The involvement of local community still remain limited other than for the traders, transport operators and hoteliers.

However, some welcome changes are coming up especially in the organized sector. The trekking trails laid by the Government of Sikkim or ecotourism promoted by West Bengal Forest Development Corporation may be cited as good examples. Ecotourism is also being promoted by some NGOs – for example 'Help Tourism' in Sundarbans, West Bengal and in Northeast India.

The operational guidelines in the National Tourism Policy, 1997 mainly refer to management plan, eco-friendly structures, monitoring of adverse impacts etc as also incentives for ecotourism operators, provision of visitors information, preparation and distribution of codes of conduct.

Sadly, much of the above guidelines remain in paper!!

3.10 Waste Management

India over last two decades has issued legal directives with regard to handling of Hazardous Waste (1989), Biomedical Waste (1998), Municipal Solid Waste (2000) etc. There is no specific directive for disposal of e-Waste. The mode of disposal of nuclear waste has never been in public domain.

The legal instruments however have hardly been obeyed. Most of the states failed to create designated disposal sites for hazardous waste and as a result fate of such waste material remain fuzzy. Medical establishments across the country both run by the Government and Public Sector have been alerted but except for four metropolises, there is no 'Action taken' report. Medical waste including surgical cotton, used needles and syringes, sample vials for pathological testing find way to an unorganized market. Even in four major metropolises, percentage of medical establishments complying with the directive has never been made public. Even from government run hospitals such waste material is pilfered.

The Municipal Solid Waste (Management and Handling) Rules 2000 promulgated by the Government of India under section 3, 6, 25 of the Environment Protection Act, 1986 provided detailed guidelines for anaerobic digestion, composting, land-filling, recycling, segregation, transportation etc. The rules however remain in law book for most of the municipal authorities across the country even after 6 years of its publication. The State Governments have the duty to enforce the provisions of these rules and State Pollution Control Boards have been given the power to monitor compliance of standards, grant authorization for waste processing and disposal facility and coordinate with CPCB for review of the standards and compilation of data. But in spite of all the specific directive, the rules remain without compliance.

The State Boards have also been given the power to enforce 'The Plastic Manufacture, Sale and Usage Rules, 1999'. In spite of the power to grant or renew registration for manufacturers a market study will reveal sale, use and free distribution of plastic carry bags with thickness below the prescribed limit of 20 m. Manufacturers therefore pay little heed to the rules and the State Boards ignore the process of violation.

3.11 Human Settlement

The Human Settlement and Environment is interlinked with the issue of quality of life. The Policy Statement, 1992 pointed out to the need of gainful employment and provision of basic needs like communication, medical, educational facilities and improvement of infrastructural facilities like water supply, sewerage, solid waste disposal and transportation in an integrated manner. The policy also focused on the needs for open space, green cover and called for environmental appraisal of projects related to urban development. It touched on the environmental health problem and need for monitoring epidemiological data to ensure adequate early warning system for prevention and control of diseases. Out of the 17 bullet points listed in the future strategy only a selected few have been focused above.

Now a look at the current situation may reveal the progress since India's initiation in the field of Environment in 1974 or publication of the Policy and Strategy Document of 1992. The State of India's Environment – The Citizens Fifth Report by the Centre for Science and Environment (1999) stated 'Almost every Indian Small and Medium town is on its way to becoming an urban nightmare......too much traffic, mushrooming slums, piles of garbage and high levels and water and air pollution are making life hells for residents of these towns.... The worst hit are the poor, who are often jobless and have to scrounge around for food and shelter'. One may argue within a span of seven years (1992-1999) no miracle can happen. Sadly the picture remains grim even after 14 years in 2006.

Industrial towns are allegedly crushed by their own growth. Starting from Ludhiana in the west to Durgapur – Haldia in the east, the scenario remains the same.

Most of large urban metropolises suffer from shortage of water, ever increasing vehicle population (0.306 million in 1950-51 to 34,216 million in 1995-96) and consequent auto emission, lack of waste disposal system except in selected high profile sites inhabited by the rich and the famous of government institutions; lack of green cover, open space, parks and garden is visible all over except in the capital city of New Delhi; so is plastic pollution. One need not cite statistical figure. Examples of unplanned growth was evidenced in a recent ICC-EU programme which presented an excellent Environmental Risk Analysis System in a recent meet at Calcutta for the emerging chemical hub in Haldia, West Bengal. It was shocking when the Head of Haldia Petrochemicals disclosed that the township lacks a Trauma Centre and well equipped hospital! So no amount of digitized software can save in case of a real life crisis! Lesson of Bhopal is still being ignored.

India has grown from 76 class I towns to 296 in 40 years (1951-1991), 91 to 341 class II towns, 327 to 927 class III towns, 608 to 1135 class IV towns and declined from1124 to 725 class V towns, 569 to 185 in class VI towns. But such growth did not help the poor; slum population still makes up a significant part of urban profile (± 25.5 per cent only in major cities). It took 20 years to come down from 28.7 per cent to 25.43 per cent (between 1981 – 2001) in 23 class I cities.

Access to the safe drinking water data shows 81.38 per cent of urban households and 55.54 per cent of rural households have such facilities, the urban – rural combined percentage stands at 62.30 (CSE, 1999, Part II, p.121). But the question remains how safe is safe? None of the class metropolises guarantee safe drinking water; tap end analysis of supplied water is almost absent. Maximum urban population suffers from water and air borne pollutants. Rural population still remains single largest target of enteric diseases due to lack of access to safe drinking water.

Sanitation is available to only 23.70 per cent of Indian population. The figure is shocking in case of rural population – it stands at less than 10 per cent after 50 years of India's independence (CSE, 1999, Part II, p. 124).

The situation regarding sewage system in urban centers remains equally poor with less than 50 per cent being served by any such services.

No consolidated data on open space, parks and garden and the extent of green cover in urban centers of India is available.

5.0 DISCUSSION

The foregoing analysis of progress based on data, vis-à-vis the basic priorities and strategies for action (Population Control, Conservation of Natural Resources) and also development policies from environmental perspectives (Agriculture, Forestry, Industries, Mining, Tourism, Human Settlement) clearly indicate the current situation. In spite of capacity building, laws and standards, institutional and financial support grading in the progress report can at most be ranked as "fair". The reasons of failures may be probed and listed. Absence of political willingness and lack of administrative strictness for implementing the strategy may be the single most cause for such a dismal profile. Corruption in the civil administration undoubtedly contributes further to such a situation.

An opinion poll of citizens about the annual environmental quality reporting may reveal the verdict of the civil society. While Judiciary in India has become more activated on environmental issues, violation of legal/ verdict continues unabated.

The NGO's in India have emerged as an identifiable community. The Government of India stressed on NGO participation in environment conservation and management since setting up the Department of Environment in 1980 but the effective collaboration between the civil society organization and system of governance deserves much to be done. NGOs are mostly asked to generate data or report public complaints but hardly become a part of action – plan. So basically the role become peripheral.

Further, since opening up of market in 1991, at least 18 new policies concerning Population, Agriculture, Water, Health care and Environment etc., have been announced. A careful study of these policies would show a scaling down tendency in imposing environmental standards and control system. Demand for development reigns supreme while environmental concerns are being soft-pedaled. Corporate social responsibility for environment related issues are yet to become visible. The environmental factors in other sectors like Irrigation, Energy, Transport appear equally being placed on a "low priority". In the ultimate analysis, India's challenge to become a real "developed country" from the current status of being "developing country" should not be judged by GDP growth only; it can really be termed so when India ranks at least within 100 positions in the Human Development Report of UNDP. The achieve an upward ranking amongst 177 countries seems to be a formidable task ahead in 21st Century.

REFERENCES

Banerjee, Ajit, K., 2006. Impact of Joint Forest Management in the livelihood of the People of South West Bengal. 1-28

Baumel, William, J., and Oater, Wallace, E., 1979. Economics, Environmental Policy and Quality of Life. Prentice Hall, N.J., 1-377.

CSE, 1999. State of India's Environment. The Citizens' Fifth Report. Part II – Statistical Database. 1-256. Centre for Science and Environment, New Delhi.

DoEn, 1984. Indira Gandhi on Environment. Department of Environment, Government of India, 1-123.

DoEn, 1999. A Report on Status of Environment in West Bengal. Department of Environment, Government of West Bengal, 1-270.

Ghosh, A. K., 2007. Environment: Issues and Concern. APH, Delhi. 1-267.

Khoshoo, T. N., 1986. Environmental Priorities in India and Sustainable Development. Indian Science Congress Association, New Delhi, 1-224. s

Ministry of Health and Family Welfare, 2000. National Population Policy. Government of India.

MoEF, 1992. National Conservation Strategy and Policy Statement on Environment and Development. Ministry of Environment and Forests, Government of India, New Delhi, 1-36.

MoEF, 2003. Annual Report 2002-03. Ministry of Environment and Forests, Government of India, New Delhi, 1-259.

MoEF, 2004. Annual Report 2003-04. Ministry of Environment and Forests, Government of India, New Delhi, 1-295.

MoEF, 2005. Annual Report 2004-05. Ministry of Environment and Forests, Government of India, New Delhi, 1-177.

MoEF, 2006. National Environment Policy 2006. Ministry of Environment and Forests, Government of India, New Delhi, 1-52.

MoEF, 2006. Annual Report 2005-06. Ministry of Environment and Forests, Government of India, New Delhi, 1-286.

NBA, 2004. The Biodiversity Act 2002 and Rules 2004. National Biodiversity Authority, Chennai. 1-57.

Swaminathan, M. S., 2004. The Ever Green Revolution. Anniversary Lecture 2004. Shastri Indo Canadian Institute, New Delhi, 1-31.

Visaria, Pravin, 2002. Population Policy: Beyond Number. Seminar, 511: 14-28.

SUBJECT INDEX

Agenda 21 5,22

Agriculture 121

Air Pollution 208

Biodiversity 61,70,85

Biomass 164

Biodiesel 182,184,187,194

Biopesticide 198,199

Biocontrol 199

Biofuel 181,190,192

Club of Rome 2

Chipko 23

Climate Change 61,79

DVC 56

Energy 151
    renewable 154

Ethanol 183,185,189,193

Forestry 124

EPC 24

GDP 143

GHG 113

Glaciers 121

Globalization 22

Global Warming 112,137,176

Health 128

Human Rights 20

Industry 131

Jatropha 181,185,186,188,189

JFM 24

JVA 56

Land 34,42
    degradation 34
    policy 34,39

Megadiversity 74

Millennium Development Goals
    7,14,16

Nuclear Power 166,170

Pesticide 202

Pongamia 181

Population 1,3,4,13,76,77,79,141

Poverty 10,11,13,16

RTI 28,29,30

Sacred Groves 84

Species
    invasive 99,100,103

Silent Valley 23

Solar Power 158,159

Sustainable Development 7,15

TRIPS 22

UNCED 5,22

Urban Greening 206

Water 53

Watershed
- urban 54,56

Wildlife 92
- tiger 93
- tourism 97
- trade 97

Wind Power 161

WTO 12